D1156602

Man and Water

By the same author

VICTORIAN TECHNOLOGY

THE HERITAGE OF SPANISH DAMS

A HISTORY OF DAMS

MAN AND WATER

A History of Hydro-Technology

Norman Smith

CHARLES SCRIBNER'S SONS

1 3 5 7 9 11 13 15 17 19 I/C 20 18 16 14 12 10 8 6 4 2

Printed in Great Britain
Library of Congress Catalog Card Number 75-24865
ISBN 0-684-14522-7

FOR MY PARENTS

Contents

Illustrations

(*Between pages* 114 *and* 115)

Diagrams

Fig.

Preface

Man's relationship with water has been vital, complex and varied from the earliest times. The evolution of hydraulic technology has been stimulated and fostered by questions of convenience, utility and necessity. By and large this book is about necessity. It is concerned with society's fundamental need to master the basic problems of food production, the supply of potable water and the generation of power. That is how the material is presented, in three parts, and this division is as appropriate to a historical study as it is to a modern one bearing on a trio of issues which are of profound significance to the present day development, management and utilization of a vital and precious resource. The history of man's use of water shows that great changes have occurred; it also proves that nothing has changed.

Some topics in this book have been written about before; others have never been presented. Certainly nothing so comprehensive has been attempted to date and there is no denying that in striving to be so comprehensive a high degree of selection and sustained compression have been unavoidable; there were no options so long as a broad and synoptic view was the aim. I hope this aim has been realized and that the necessary selectivity has not led to serious omissions or any problems in comprehension.

A word about units. For the most part English units have been used but where the sources quote metric dimensions or weights these have usually been retained. This ensures that measurements which are approximate or rounded in the original are not given the *appearance* of precision by being converted.

I would like to record my thanks to a number of people who have contributed to my efforts. Laura Brown typed most of the manuscript; Janet Reader finished it and then coped patiently and exactly with a host of tedious alterations and corrections; my old friend Tony Kerr found time in his busy routine to manufacture illustrations; John O'Leary printed Plate 8.

Once again it has been a pleasure to work with Peter Davies and I would like to thank Derek Priestley for his help and advice and numerous suggestions for improving the text, nearly all of which I was happy to adopt.

Finally, for encouraging and supporting my endeavours in research and authorship, a special word of thanks to Professor and Mrs A. R. Hall.

Norman Smith
Imperial College, London
December 1974

Part I

Water and the Land

I

Ancient Irrigation

THE DETAILS OF the argument vary, in places considerably, but in general there is agreement that in antiquity irrigation was a key technology, not only in itself but also as an influence in the emergence and development of other technologies, and above all in its effect on society and the social order.[1]

Man's change from a food-gathering, animal-hunting, nomadic existence to a settled way of life based on food growing and the domestication of animals was a momentous, complex and very long-term process which is still anything but well understood. While this change cannot be said completely to define man's evolution from a primitive to a civilized way of life, it was certainly a focal point of this important shift in humanity's status. The techniques which enabled the change to occur were many and various – tool-making, simple metallurgy, the use of fire, some agricultural know-how – and generally speaking a feature of the process was a slow migration of peoples to the valleys of big rivers.

The concept of ancient river valley civilizations has been widely accepted for many years[2] and the five valley societies of importance are well known: Egypt in the Nile valley; Mesopotamia in the Tigris–Euphrates valley; Indian civilization in the Indus valley; China in the Yellow river valley; and Andean civilization in river valleys of coastal Peru. To what extent these groups had practised, and to a degree even perfected, irrigation technology before they took on the mighty problems posed by the Nile or the Tigris–Euphrates is difficult to discern and impossible to generalize about. Some small-scale irrigation, perhaps on the tributaries of a big river system, is plausible. However, this is not to say that a developed irrigation technology was a pre-requisite for the emergence of river valley civilizations. It did become, however, a fundamental and critical component of their evolution and growth.

We should perhaps notice two other features of irrigation and its interrelation with the initial emergence of civilization. River valleys, with or without irrigation technology, are not essential to the development of civilized society. It can and has happened elsewhere, for instance in the cases of Crete, Middle America and Ceylon. On the other hand, it was not only in river valleys in ancient times that irrigation was practised. Irrigation sometimes developed in other situations and, specially adapted, it was the basis of achievements just as notable as those found in river valleys.

Although certain features are common to all irrigation techniques, nevertheless variations in terrain, river régimes, crops and climate produce distinctive systems in different places. And this was true of antiquity. Indeed, one of the most fascinating aspects of ancient irrigation was the capacity of engineers to evolve methods which were so well adapted to local needs and conditions.

In Egypt everything to do with irrigation, and a good deal else besides, was dictated and conditioned by the Nile, and one factor was of overriding importance: once a year the river floods, regularly and predictably, beginning in the middle of August, reaching a peak at the end of that month, and receding two months later. A famous feature of the Nile flood, noticeable to ancient and modern eyes alike but differently interpreted, is its colour, a deep reddish-brown imparted by the large amounts of silt scoured from the mountains of Ethiopia. It is this sediment which over the millennia has raised the Nile's bed, formed a delta in typical fashion and, until very recently, fertilized Egypt's agricultural land. The earliest Egyptian use of fertilization by flooding was uncontrolled. Man merely planted in the fertile swamps left behind as the flood naturally subsided. But as Egyptian society developed and its demands for food grew, so agriculturalists learned to modify the natural process. In essence it was necessary to exert more control over a larger irrigated area. And the problem of river control meant more than an extended use of flood water for irrigation. It was vital too that massive irrigation works did not afford the Nile, in fuller spate than usual, an opportunity to rampage through fields and towns alike and so destroy the very thing it sustained. 'Egypt,' as Herodotus so aptly remarked, 'is the gift of the Nile.'

Basin irrigation was the technique which Egyptian irrigationists developed. Along the river's banks, connected to the river and each other by canals, large plots of land were provided, suitably levelled and at a correct level in relation to each other. Every season flood water was run on to the land by breaking dykes, it was allowed to stand while the fertile silt settled, and then the excess water was drained back into the river when the inundation was receding. This was the whole basis of Egyptian irrigation until the nineteenth century when an element of perennial irrigation using dams was introduced for the first time. In ancient Egypt dams across the Nile were not built, although occasionally quite large feeder canals were laid out to convey the river's flood water to points somewhat beyond the basin system, to the Fayum depression for example.

Basin irrigation was ideal in the Nile valley. Because the river is relatively steeply graded good drainage was obtained, the canal system was short and kept silt-free by the ready flow of water which also helped to flush out salts. There were disadvantages, however. Obviously only one harvest was possible per year and the system required a lot of water.

If the nature and history of Egyptian irrigation are both well understood, the same is by no means true of Mesopotamia. In virtually every way the picture is more complex, and central to this situation is not one river but two. After a turbulent passage over the hundreds of miles of their upper reaches both the Tigris and Euphrates suddenly find themselves in new terrain when they hit the huge flood plain which now covers most of modern Iraq and used to be Mesopotamia. The 'two rivers' are not like the Nile; their shallow gradient has encouraged both to meander, sometimes on a massive scale, and both are liable to large and sudden flooding. Curiously though, both these features may well have stimulated ancient irrigation. The Euphrates is generally at a higher level than its larger but more docile neighbour so that flood channels formed by the Euphrates on its eastern bank occasionally found their way to the Tigris. Conceivably these naturally formed channels suggested the essence of the technique which was to become the most distinctive feature of Mesopotamia's irrigation. Using the Euphrates as a source, water was led across country through a network of canals which finally drained into the lower-level River Tigris.

Although the basic features of Mesopotamian irrigation are evident enough, the precise nature of the engineering works involved is impossible to determine. A lack of surviving technical records (and probably few ever existed), extensive development by later peoples, some radical changes in the rivers' courses, and six centuries of desolation and neglect since the Mongol invasions, have conspired to obliterate all the details. It *is* clear, however, that Mesopotamia unlike Egypt was a land of dams.[3] Perennial irrigation requires reliable flow control between the primary canals and the matrix of secondary canals they feed. For the most part the dams, sluices and regulators of Mesopotamia must have been small structures but at canal headworks quite substantial construction was required. Details of the type of work which was being undertaken in the seventh century B.C. can be deduced from the archaeological remains of the irrigation and water-supply works laid out around Nineveh by the Assyrian king Sennacherib (see Chapter 6). Subsequently, the Achaemenian Persians elaborated and extended Mesopotamia's irrigation into the system which so impressed Alexander the Great and was later developed by Sassanian engineers.

As its name denotes, perennial irrigation has the advantage that more than one crop can be grown per year. But in Mesopotamia the price to be paid was high, and in the end too high. The quantities of silt carried by the rivers and eroded from the irrigation channels themselves were for ever choking a canal system which was very long. And the tendency of the river system to sudden and heavy flooding was a constant and serious threat to both river and canal banks. Overall then, the maintenance of Mesopotamia's irrigation system was a massive problem and in the end, in Islamic times, it became overwhelming.[4] As we shall see later in this chapter the fate of the region's agriculture was finally sealed by the most deadly enemy of all – salination.

Basin and perennial irrigation are the two basic variants associated with farming in the valleys of big rivers. Two other techniques appropriate to other situations complete the early repertoire.

The origins and development of terrace irrigation are obscure. But it was a very basic technique, used at an early date in Syria and Palestine, and frequently encountered in India, China and pre-Columbian America. The last locale is important since it indicates that the idea did not have to diffuse from some single point of origin. Terrace irrigation is fundamentally a method used in hilly country and requires the formation of an elaborate series of terraces stepped down a hill-side. Water is run from one level to the next in channels. The expenditure of effort involved in preparation and maintenance is high in relation to levels of production. It is in the nature of terrace irrigation that rivers cannot usually furnish a supply of water. Instead use must be made of artificially stored rainfall, wells, springs and occasionally, when conditions are favourable, qanats (see Chapter 6).

Finally, in the case of wadi irrigation, our last category, the problem of water resources is the most critical of all. Indeed it is amazing that ancient irrigation engineers, in a few instances, succeeded in watering quite large areas with only the most meagre of water supplies to call upon. Leaving aside numerous small scale efforts, the two most striking achievements were in Yemen and the Negev. Sabaean society flourished in south-west Arabia for the better part of the first millennium B.C. Agriculture around the capital city, Marib, was sustained by the rainstorms which occasionally inundate the high mountains of western Yemen and feed water to the Wadi Dhana. Across this watercourse, some 3 miles above Marib, one of the biggest of all ancient dams was constructed. It was an earthen structure 2,000 feet long, and following its original construction in the eighth century B.C. it was successively raised to a final height of 14 metres. This massive structure was never used to impound water; it served only to lift the wadi floods to increasingly higher levels in order to irrigate more and more land by means of a canal system which used the wadi itself as a drain.[5]

The Marib example represents successful wadi irrigation using a single, big dam. The technique evolved in the Negev by the Nabataeans went to the other extreme and utilized thousands of little barriers strategically sited across one wadi after another in order to divert or capture the one or two weeks of run-off occurring each year. The system contained strong elements of the terrace technique because the passage of silt-laden water gradually filled the 'reservoirs' and stepped off each wadi into a terrace-like system. The Nabataeans' industry in building up and maintaining such irrigation works was remarkable, and the prosperity of the Negev unequalled until modern times.[6]

Although the adjectives 'sophisticated' and 'elaborate' might well be applied to many an ancient irrigation system the technology itself was neither sophisticated nor elaborate. After all, the basic proposition involves the capture of water, channelling it on to the land, and taking care of the drainage of the surplus. The man who gets buckets of water from a river and waters plants with them is an irrigator. The immense technical, social, political and cultural importance of ancient irrigation stems from the huge scale on which this fundamentally simple operation was practised.

Among a variety of major technological consequences of ancient irrigation, the development of numerous facets of civil engineering was of fundamental importance. The construction of dams and canals, matters to do with water flow and control, and elaborate surveying problems, all presented themselves uncompromisingly. Their successful solution was in the hands of experts. Whatever the name by which they were then known, e.g. 'Superintendent of the Irrigated Lands of the King', these were hydraulic engineers and their skills were not infrequently called into use for works other than irrigation.

Not all irrigation needs could be met by channelling water through a canal system. Frequently it was necessary to lift water from one level to another and for this purpose irrigators introduced at least one mechanical device at a very early date. The famous shaduf, sometimes called swape, is a lever device of great antiquity (see Figure 1); it is pictured as early as 2500 B.C. in Akkadian reliefs and *c.* 2000 B.C. in Egypt. The shaduf is not a machine in the sense that it replaces the effort of men or animals. But it does enable a man or a team of men to work more smoothly and faster. The shaduf has remained in the repertoire of simple irrigation devices to the present day and its application is world-wide, being not unknown in Poland, Canada and Scandinavia.

The first use of animals to raise irrigation water, a practice which ultimately became, and still is, commonplace, is very obscure. The first domestication of the camel and ox, which occurred at different times in different places, must surely have been applied to the laborious job of water-raising quite soon, although probably the equipment was simple, teams of animals marching down ramps and drawing ropes over pulleys. The development of animal-driven wheels, indeed water-wheels generally, is quite late, Hellenistic at the earliest.

Irrigation acted as a spur to the development of certain aspects of astronomy and mathematics. The need was particularly acute in Egypt, since the functioning of a whole year's irrigation depended critically on an accurate prediction of the onset of the river's annual flood. The rising of Sirius was the astronomical event which became associated with the beginnings of the inundation and together both events may have served to show that the Egyptian civil calendar of 360 days was 5 days short. Similar relations between the positions of the stars and agricultural cycles evolved in Mesopotamia, although here the question of flood cycles was much less central.[7]

The role of mathematics in irrigation was fundamentally a matter of

Figure 1 The ancient shaduf as depicted by Thomas Ewbank in his *Hydraulic and Other Machines* of 1870.

mensuration. Land surveying, for instance, involved more than just the question of levelling, although that was crucial enough. Land areas also needed to be assessed quantitatively in order to calculate taxes and crop yields. Actually some of the oldest of all technical drawings are Mesopotamian plans, 'drawn' of course on tablets, of canals and fields. It is by no means clear, however, whether the principal use of such drawings was technical or administrative. Volumetric measurement of both soil and water and also quantities of agricultural produce were undertaken, and here and there one finds examples of the calculation of work quotas and rates of payment of labourers. While acknowledging the important role of irrigation – and building should be included also – in fostering some early uses of mathematics, it should not be concluded that the arithmetic involved was anything but extremely simple. The measurement of areas was limited to operations with squares, rectangles and right-angled triangles, and volumetric measurement went little further. We may be pretty sure, too, that whatever instrumentation was used in the course of practical measuring exercises, it was elementary in both operation and construction.

While the remarks which open this chapter admit the difficulty of determining the extent to which irrigation played a part in the formation of the earliest settled communities, there is no question that once such groups were established, irrigation problems were central to the evolution of the social structure and cultural forms of ancient societies. Such well worn terms as 'fertile crescent', 'river civilizations' and 'irrigation civilizations' are acknowledgement of the assumption. A recent focal point of this whole discussion has been Karl Wittfogel's fascinating and provocative book *Oriental Despotism*, in which ideas about hydraulic society and its influence have been taken further than ever before.[8]

It would be hopelessly ambitious to attempt a summary of the whole of Wittfogel's complex and wide-ranging thesis, perhaps one should say theses, but the essential points for our purposes are these: the supply, control and use of water on a large scale, (the scale of Egypt, Mesopotamia, the Indus, China or Peru for example), is bound to produce great technical, organizational and administrative problems. These can only be met and solved by strong and stable government, powerful in its centralized control of money and manpower. Wittfogel argues, therefore, that large-scale irrigation must be the outcome of such authoritative and determined bureaucratic government. His view has been challenged,[9] vigorously on occasions, on the grounds that in reality huge irrigation systems were not conceived in their fully developed form and then laid down in their entirety in a short period of time. Instead, the critics point out, the systems developed and grew by small degrees, in increments which financially, constructionally and organizationally were well within the resources of groups falling far short of Wittfogel's mighty despotisms. All the same, however a large irrigation system is built up and no matter how long its formation may take, there results in the end a massive problem of administration, repair, maintenance and so forth. Wittfogel's concept of strong government with control of manpower, his idea of hydraulic society, ought now to operate.

Actually there is evidence that it need not, that discrete elements of what comprises an irrigation system which is large overall, may well be run as quite small autonomous units. For instance, if an irrigation system is based on a number of different rivers acting as sources, then it may well be expedient for each to be operated as a separate irrigation unit. It is, after all, a notion which is current at the present time. The sweeping scale of Wittfogel's work is such that one is bound to be able to contest some aspects of most of it. But the core of the thesis contains much that is very persuasive. The historical record does suggest that between large-scale public works (irrigation, water-supply, road systems) on the one hand, and strong stable government functioning in times of peace on the other, there is a correlation. To put the point another way, large-scale public works have yet to be shown to be the achievement of small, disorganized, uninfluential societies, while civilizations of any importance in China or India, the Middle East or America, generally produced public works of some description.

The importance of irrigation in antiquity as an influence on government and social organization can be illuminated by a number of issues. The city-states of ancient Sumeria were basically irrigation units or provinces – Drucker actually calls them 'irrigation cities'[10] – and a frequent cause of their coming to blows was irrigation problems. This is in line with another feature of Drucker's argument, namely that irrigation cities gave rise to the first standing armies because of the need to protect the defenceless farmer and his extremely vulnerable land. As succeeding dynasties developed and extended Mesopotamia's irrigation, so there are references

If a man neglect to strengthen his dyke and do not strengthen it, and a break be made in his dyke and the water carry away the farm-land, the man in whose dyke the break has been made shall restore the grain which he has damaged.

If he be not able to restore the grain, they shall sell him and his goods and the farmers whose grain the water has carried away shall share (the results of the sale).

If a man open his canal for irrigation and neglect it and the water carry away an adjacent field, he shall measure out grain on the basis of the adjacent fields.

If a man open up the water and the water carry away the improvements of an adjacent field, he shall measure out ten GUR of grain per GAN.

Figure 2 Four sections of Hammurabi's Code dealing with questions of irrigation and the use and misuse of water.

to the provision of labour for canal construction, to contracts, and occasionally to the legal framework which governed the use and abuse of land, water and equipment. The prime example, by far, of the last is Hammurabi's Code, *c.* 1800 B.C., whose references to irrigation are direct, stringent and unambiguous. They are set out in Figure 2.

Egypt had no formal legal code comparable to Hammurabi's but in other respects irrigation was comparably important. One finds, for instance, that the old hieroglyphic sign for 'province' is a pictograph of an irrigation unit. Such provinces were administered and staffed by a variety of officials bearing such names as 'inspector of the dykes', 'chief of the canal workmen', 'watchers of the Nilometers' and so on. To deal with arguments there were special water tribunals, a type of institution which has a history lasting long after antiquity and turning up in many places.

It is generally believed that the labour which was regularly needed to prepare and maintain dykes, canals and fields was a form of corvée. The care of the crops, however, rested with a few farmers. Thus irrigation not only formed the basis of a tax structure but it also put the provision of food in the hands of a few, leaving the bulk of society free to soldier, to trade, to administer, to pursue science and the arts and to cultivate religion. Indeed, in some respects, religion was closely related to irrigation itself. Big rivers like the Nile, Tigris and Euphrates figured prominently in the legends, myths and cults of the respective civilizations. Key events in the irrigation calendar were marked by religious festivals and frequently were graced by the presence of the priestly hierarchy. The maintenance of proper relations with the 'river gods' was crucial since it was generally accepted that a bad harvest was an unmistakable sign of divine retribution. And the most savage retribution of all was floods, the great Biblical flood being an exaggerated and telescoped rendering of the many flood legends, doubtless based on real disasters, which figure in ancient Mesopotamian stories.

If large irrigation schemes depended for their construction, development and operation on the organizing role of stable government, we may expect to find their demise associated with administrative decay. Sure enough there are examples – though some of the best instances are not in antiquity – but generally other factors were at the least influential, and occasionally critical. The extensive gardens on the banks of the Wadi Dhana at Marib blossomed so long as the Sabaeans prospered from the traffic of goods from India and China moving by caravan to Egypt, Syria and Mesopotamia and also from the local production of spices and incense. Two things conspired to destroy the economic base of Sabaean life. The rise of Christianity largely killed the market for incense while the influence of Rome in matters of trade ensured that the caravan route up the Arabian peninsula was replaced by a predominantly maritime connection using the full length of the Red Sea as far as Egypt. Marib's dam and canals, from local inscriptions evidently a difficult layout to maintain at the best of times, fell into ruins as soon as times were bad. Attempts at revitalization, the last in the sixth century, all failed and to this day the area has never recaptured its fertile past.

It is certain that the irrigation traditions of the Sabaeans exerted no influence on the development of Roman agriculture; only once did a Roman army penetrate southern Arabia and the result was a disastrous retreat. The extent to which the long established techniques of Mesopotamia influenced Roman engineers is impossible to determine. Probably the influence was very little. The Euphrates frontier was remote and inhospitable, a region to be defended vigorously when necessary (and by the third and fourth centuries that was pretty often), but otherwise hardly an area of close contact.

In other places things were different. Egypt was established as 'the granary of Rome' at an early date and in the Negev region the idea of conserving soil and water behind wadi dams was rapidly assimilated following the conquest by Trajan in A.D. 106. A wholesale application of Nabataean techniques along the north of Africa served to sustain life in magnificent cities such as Sabratha and Leptis Magna. Special adaptation to the climate and topography produced the most massive of all irrigation schemes of wholly Roman origin.[11] To minimize flood damage, to conserve and direct water and to exercise a degree of soil conservation, large dams of masonry, or earth faced with masonry, were erected across wadis much greater in width and prone to much heavier flows than those of the Negev.

Elsewhere, the Romans absorbed irrigation technology from other sources. In Italy itself, there were the existing efforts of the Etruscans to the north, and here and there Greek colonists to the south. The Romans were no mean farmers as an extensive literature, including such famous authors as Cato, Varro, Columella and Palladius, indicates.[12] In Italy irrigation was not for the most part a critical problem, although some crops in certain places benefited considerably from it. Further west in Spain, especially in the coastal areas of Phoenician and Greek settlement, irrigation must in view of the climate have been practised in greater earnest. All the same, a coherent picture of Roman irrigation in Iberia has yet to be drawn and the lack of one represents not only a gap in our view of Roman technology, but quite seriously undermines efforts to understand later developments, especially those of Moslem origin.

Because the quantities of water involved were so large, irrigation was the key stimulus in antiquity to the development of water-raising machines. By Roman times the evolution of the basic repertoire was well advanced as Book X, Chapter IV of Vitruvius's *Ten Books on Architecture*, written about 27 B.C., shows conclusively.

A. G. Drachmann[13] has suggested that the Archimedean screw (see Figure 3) is probably justly named – hence its origins are dateable to about 250 B.C. – and by Roman times it was widely used. Writers in the first century B.C. mention its application to irrigation, mine drainage and bailing ships. Vitruvius gives a matter of fact description with clear instructions on how to make one. A painting in Pompeii shows an Archimedean screw being trodden by a slave.

For raising large quantities of water and for all high lifts – ten feet or more – other machines were favoured: the tympanum, the noria, and best

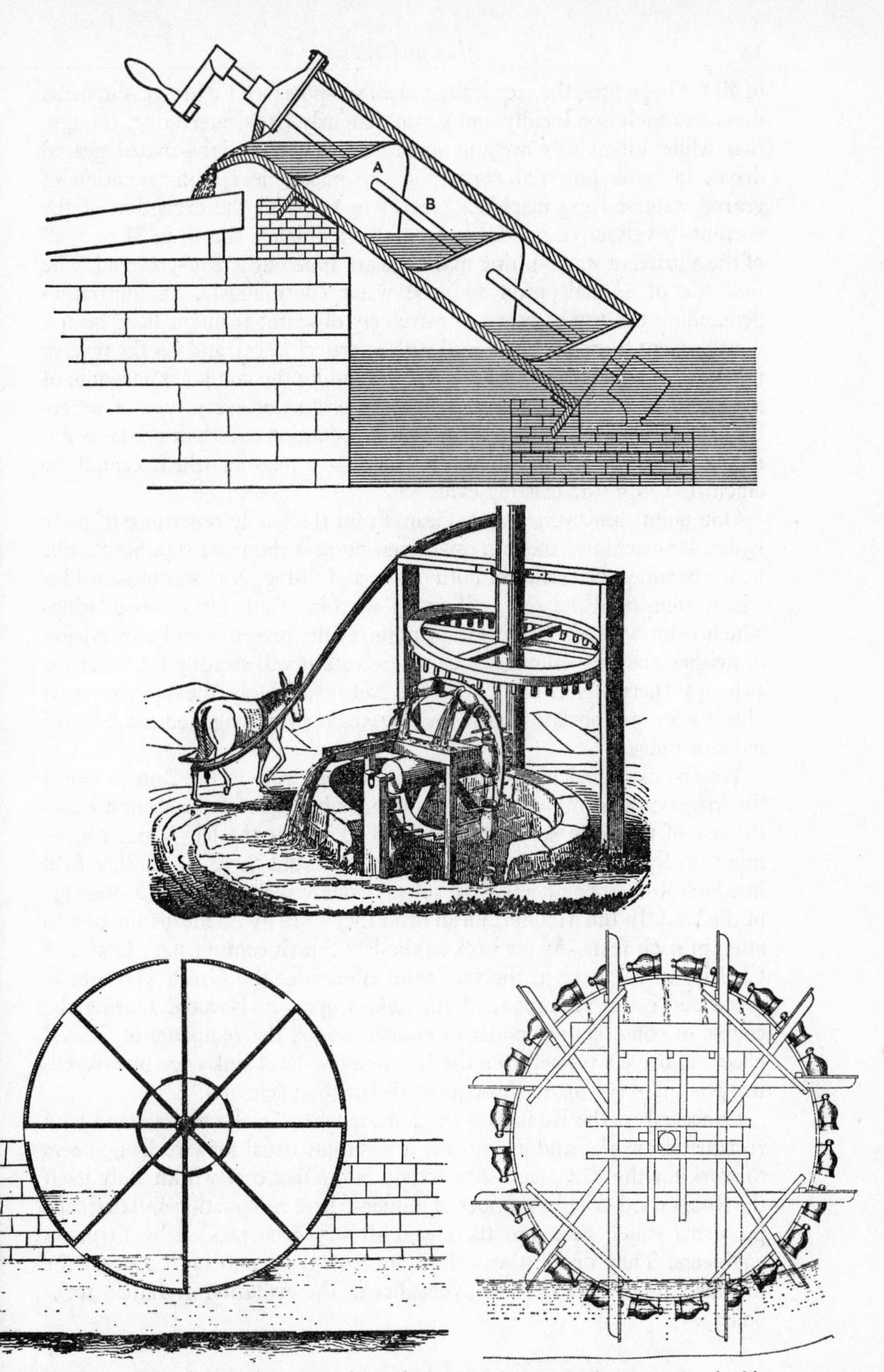

Figure 3 The four basic water-raising devices used in antiquity: *top*, the Archimedean screw; *centre*, the saquiyah or chain-of-pots; *bottom left*, the tympaunm; and *bottom right*, the noria.

of all for large lifts, the saquiyah or chain-of-pots (see Figure 3). Vitruvius describes each one lucidly and unambiguously. It is interesting, though, that while Vitruvius's account includes the use of right-angled geared drives in water-powered corn mills, he makes no specific mention of geared water-raising machines (see Plate 1). With the exception of the current-driven noria, a modern example of which is shown in Plate 2, all of the Vitruvian water-lifting machines are apparently man-powered. The first use of animal-power to raise water continuously, an innovation demanding the use of gears, is extremely obscure; it might have been a development from the water-mill with a vertical wheel, and yet the reverse influence is equally plausible. Notwithstanding the confident accounts of a number of writers, the fact is that the origins of every type of water-lifting machine, the sequence of their introduction and their relationship, if any, to developments in water-power, are matters which cannot be elucidated from the existing evidence.

One point, however, is very clear. From the whole repertoire of early hydraulic machines, the current-driven noria is the most significant. For here, spanning the realms of both power and lifting, is a most elegant idea which comprises the first self-acting machine (but *not* self-regulating) which man evolved. Its operation requires the presence and supervision of neither man nor animal. Once set to work it will steadily lift water for as long as there is river flow to drive it. Not surprisingly, the noria has been able, under appropriate conditions, to remain a key irrigation machine till modern times.

Water-raising machines, in addition to their role in the Roman world for irrigation and mine drainage (a famous example of the second being the use of norias in series to dry the workings of the Rio Tinto copper mines in Spain), also found application, here and there, in another field in which Roman engineers were active, namely reclamation and drainage of the land. In the Mediterranean orbit they were by no means the first to attempt such feats. As far back as the late fourth century B.C., Crates of Chalcis, an engineer in the service of Alexander the Great, very nearly succeeded in an attempt to drain Lake Copais in Boeotia. During the course of constructing modern drainage works, the remnants of Crates' efforts to drive a tunnel from the lake to a low-level sink were uncovered, the precedent for many subsequent reclamation schemes.[14]

Conceivably, the Romans were familiar with Greek drainage and land reclamation works, and in any case it was quite usual for Greek engineers to work for the Romans. More certain is the fact that within Italy itself the Romans inherited a number of immense land reclamation and drainage problems which here and there had already been tackled by Etruscan engineers. They divide themselves into two types and both suggest the intrinsic importance of land hydraulics in the evolution of Italian technology.[15]

Poor natural drainage, combined in many cases with the marked disposition of several large rivers to flood low-lying ground, produced conditions which were unhealthy for the inhabitants and unproductive agriculturally. Systems of drainage canals were established in many places in attempts to improve matters. Often the attempt was not unsuccessful. At Veii, near Rome, the fields north of the city were considerably improved and the extensive Campi Rosei near Rieti was first drained with canals in 271 B.C. Throughout the Po valley Etruscan influence was marked, but not notably successful. The Romans worked endlessly to tame this disagreeable river so that valuable arable land on both banks would not be periodically washed out. A complete solution to the problem was not of course possible – any more than it has been in modern times – but considerable improvements accrued, and the same can be said of schemes laid out near Ravenna and further north at Veneto. On the other hand, some land reclamation efforts must be judged a failure. Italy's western coastal plain is poorly drained from the notorious Maremma in Tuscany right down to the famous Pontine Marshes south of Rome. Some contemporary accounts claim a good deal for a succession of drainage canals built in these areas extending from the time of Appius Claudius (312 B.C.) through to Theodoric the Goth (fifth century). Frankly, these claims must be held in some doubt. Both regions are so wet, so low-lying and so near the sea that, as modern engineering has shown, pumping on a scale far beyond what the Romans could provide is the only solution.

Draining and reclaiming low-lying swamps and river-valleys was one facet of Roman land hydraulics in Italy. The other involved a number of bold tunnelling exploits designed to empty inland lakes. One of these rates as being among the most ambitious of all Roman engineering works. In A.D. 41 Claudius embarked on a massive tunnelling project designed to connect the huge inland lake called Fucinus, near modern Avezzano, with the valley of the River Liris 3½ miles away. Seutonius believed that 30,000 men worked on the tunnel for ten years while Pliny the Elder, who visited the works in progress, was as impressed by this deep tunnel and its many miles of access shafts (*putei* and *cuniculi*) as by anything he had ever seen. The first tunnel, finished in A.D. 51, was not a success due, it is surmised, to faulty inlet works which subsequently were rebuilt. For how long the tunnel then functioned is not very clear, but ultimately it fell into such disrepair that in the nineteenth century the whole scheme had to be reworked. Another drainage work similar in principle to Lake Fucinus but smaller in scale (and seemingly more successful in operation) was constructed for Lake Albanus near Rome.

Bearing in mind that Roman engineers also tackled drainage problems in France, Germany and Britain, it is evident that they invested much effort, frequently well rewarded, in this most demanding branch of civil engineering. But subsequent lack of maintenance or later reworkings, often very extensive, by other peoples are the reasons for our being less aware of these works than of more obvious attainments in bridge building or monumental architecture.

Following the collapse of the Roman Empire in the West, the tradition of Roman hydraulic engineering was carried on in the Byzantine world but not, so far as can be detected at the moment, in the sphere of land hydraulics. Of course, irrigation was practised, as it was throughout the lands of the Mediterranean, and in Visigothic Spain existing Roman techniques appear to have survived, but nothing either technically or organizationally as complete as the Roman effort was evolved again until Islamic times.

We know from a variety of Moslem historical and geographical writers that the Roman examplar in engineering was frequently noticed and often admired. And elsewhere, in Iraq for instance, there was much for the new and far-flung Islamic Empire to take over. This they did and with less disruption than is often believed, for the Moslem conquerors were more tolerant of their conquered territories than the over-used cry 'Islam or the Sword' might suggest. Hence one finds that, in the case of Egypt, the country's agriculture and irrigation were as essential to Moslems as they had been to Romans. Ultimately there was a decline, and by the twelfth century a decisive breakdown of Egypt's irrigation coincided with the final collapse of Fatimid rule. The efficient and effective use of the Nile as of old was not to be fully revived until the nineteenth century.

Moslem engineers achieved their greatest success as irrigators in the East, in and around the city of Baghdad, which was established by the Abbasid caliphate in A.D. 762 and destined to be its capital for nearly half a millennium (until 1258). The extensive irrigation system built up by a succession of previous peoples was taken over and developed even further.[16] Iraq's irrigation was at its most extensive, the agriculture at its most productive, and the region's population largest in the 'golden years' of Abbasid rule, the tenth and eleventh centuries. A huge area around Baghdad (see Figure 4) was successfully watered by five cross-country canals running from Euphrates to Tigris, water-ways large enough not only to be navigable but requiring in one case large masonry arch bridges to give access from side to side. At their heads these canals were fed by dams or side-regulators, apparently quite large structures whose failure on rare occasions was not only highly disruptive of farming but sufficiently noteworthy to merit inclusion in contemporary literature.

It was on the Tigris' eastern bank, between the river and the mountains called the Jebel Hamrin, that Abbasid engineers engaged in some of their most elaborate works. The Nahrwan canal, begun by Sassanian engineers, was enlarged, and conceivably lengthened, and also equipped with a second inlet, fed from a dam across the Tigris near Samarra. The rivers Adheim and Dyala were dammed to provide water for a huge irrigated area which finally drained into the Tigris some 150 kilometres below Baghdad. Even allowing that we can now no longer unravel all the details of dams and weirs, canals and river courses, due to the various failures of control works, neglect of the canals and the disappearance of salient features under layers of blown sand, the evidence of civil engineering in the grand manner is still manifest.

Unfortunately, it was not to last much longer. Eastern Islamic engineers

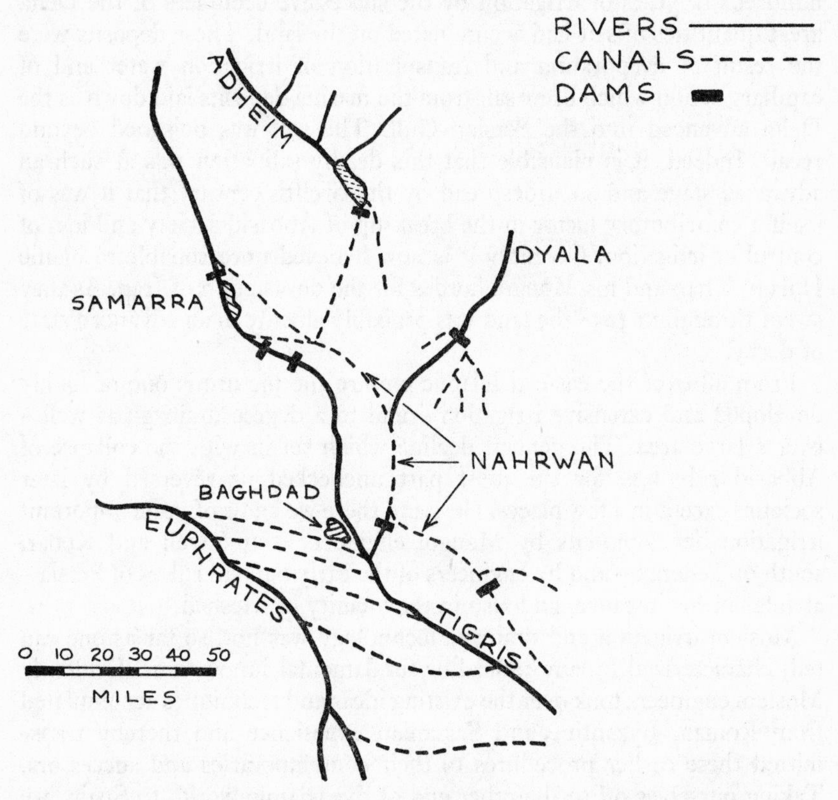

Figure 4 The system of dams, canals and rivers in the vicinity of Baghdad when the region's irrigation was at the peak of its development.

had set themselves a formidable maintenance problem which presumably, for a brief heyday, the administration in Baghdad was able to cope with by providing money and organizing labour. But as a decline of governmental power and authority began to set in late in the eleventh century, the position worsening in the twelfth century, so the canals became massively silted up and the country's irrigation quickly deteriorated. The problem was also compounded by other difficulties. Around the year 1200 the Euphrates suffered a major shift in its course south of Musaiyib, while even more disruptive was the Tigris' reversion to its former more easterly course above Baghdad. This particular change of régime – it too occurred around 1200 – was doubly disastrous, because it not only deprived land along the old course, to the west, of irrigation water, but it also wrecked the upper reaches of the Nahrwan canal to the east. Then, in addition, at about the same time, the Adheim dam appears to have failed along with some control works further south. As if all this was not troublesome enough, a more sinister malaise was advancing rapidly. After hundreds of years of irrigation by the successive occupiers of the Delta great quantities of salt had accumulated on the land. These deposits were the result of evaporation and transpiration of irrigation water and of capillary action which drew salt from the marine deposits laid down as the Delta advanced into the Persian Gulf. The soil was poisoned beyond recall. Indeed, it is plausible that this deadly salination was at such an advanced stage and so widespread by the twelfth century that it was of itself a contributory factor in the break-up of Abbasid society and loss of control of irrigation. Certainly it is now believed unreasonable to blame Hulagu Khan and his Mongol hordes for the devastation of Iraq. As they swept through in 1258 the land was probably already in an advanced state of decay.

From all over the eastern Islamic empire the picture is one of highly developed and extensive irrigation – and to a degree drainage as well – over a large area. The general decline which set in with the collapse of Abbasid rule was for the most part unchecked or reversed by later societies except in a few places. Here and there we know of quite important irrigation developments by Mongol engineers – at Saveh and Kebar, south of Teheran – and by engineers of the early Safavid rulers of Persia – at Isfahan for instance, and also in the vicinity of Meshed.[17]

Moslem irrigation and drainage technology was not, so far as one can tell, characterized by any radical or fundamental innovations. Evidently Moslem engineers took over the existing ideas and techniques accumulated from Roman, Byzantine and Sassanian experience and thereby transmitted these earlier procedures to their contemporaries and successors. Taking ourselves off to the other end of the Islamic world, to Spain, we find a particularly good example of the process at work.

2

Spain

THE MOSLEM CONQUEST of Spain was achieved with astonishing ease and rapidity. The bulk of the peninsula – a trio of northern provinces which, although raided, were never taken – was occupied in the short space of seven years (711–18). For 700 years Islam was to maintain, in the later centuries very tenuously, a foothold on the mainland of Europe. During this period, and most especially in the ninth and tenth centuries, Moslem civilization flourished in southern Spain as a tremendous contrast in many respects to the rest of the continent. That Moslem Spain was agriculturally very prosperous has been widely recognized for many years now, although details of the origin of this prosperity, the nature of its development and precisely what sort of legacy was handed on to Christian Spain after the Reconquista, continues to be debated.[1]

In the nineteenth century the majority of writers who attempted to account for the origins of Spanish irrigation were convinced that they were Islamic. Only rarely does one find any suggestion that Roman (or Visigothic) influence was responsible and that the Moslem conquerors merely copied and extended what they found already in existence. In 1864 a French writer, Maurice Aymard, in his thorough and invaluable *Irrigations du Midi de l'Espagne* was cautious in attributing *everything* to the Moslems, although he was anxious to do them justice. Other authorities followed Aymard's view and further compounded the difficulties by pointing out that whatever existed before, be it Roman, Visigothic or Moslem, a great deal must have been the later contribution of post-Reconquista Christian Spain. Land hydraulics, as we have seen, is very much an activity which evolves and extends in the hands of successive generations of irrigators and drainers; changes of régime, religion or social structure are by no means likely to bring about distinctive changes of technique or be sufficient to halt progress or developments. It must be assumed that Spain is just as good an example of this continuity of effort and purpose as Iraq or Persia in irrigation, or Italy in land reclamation.

The Romans certainly farmed in Iberia. They spread the cultivation of the olive tree, first introduced by the Phoenicians, to such effect that olive oil from Spain was being imported into Italy by about A.D. 100. A mosaic found in Ostia lists Spain as a principal corn-producing area. It is perfectly reasonable to suppose that so much agricultural production was sustained by irrigation, and very likely it followed the pattern of Roman developments in North Africa. Consistent with the overall proposition is the use in Roman Spain of water-raising machines.

What happened to Roman irrigation under Visigothic rule is a mystery. Conceivably, a good deal was kept in working order and it is evident that the Visigoths were irrigators to some degree because their book of law, the *Liber Judiciorum*, prescribed penalties for the theft of water.

So the Moslem conquerors of Spain must be supposed to have acquired a country in which patterns of irrigation were already established and the cultivation of specialized crops was traditional. To this legacy a great deal was destined to be added. It was a typical feature of Islamic expansion that men of the armies of conquest settled the newly won territories first and were later followed by large numbers of civilians. In the case of Spain the latter continued to migrate westwards long after the nominal separation of Umayyad rule in A.D. 756.

Among the Moslems who settled in Spain came large numbers of people from Syria, Egypt, Persia and Iraq and with them they brought the centuries-old irrigation techniques with which they were so familiar. The pattern of Moslem settlement was predictable. The north of the country and most of the Meseta was inhospitable to peoples from North Africa and the Middle East, and so they chose instead the river valleys of the south where the environment was familiar and manageable. That this region – al-Andalus as it was called – experienced a considerable importation of eastern agricultural techniques and irrigation practices is clear from a variety of evidence. There is, for instance, within Spanish irrigation terminology a preponderance of terms and words of unmistakably Arabic origin. Then again, there is the evidence of irrigation machinery. The whole repertoire of Middle Eastern devices – the noria, the shaduf, the saquiyah – was massively exploited in Moslem Spain. Thomas Glick has shown[2] how the noria and another highly distinctive eastern hydraulic device, the qanat, came to be characteristic only of that part of Spain dominated by Moslems. And in the long run such diffusions were to continue. Not only did Christian Spanish engineers take both machinery and qanats to the New World, but some of the water-lifting devices achieved a quite widespread distribution in the rest of Europe and in very many places continued to be used well into this century.

Crops are a third source of evidence for Moslem influence. Rice, oranges, cotton and sugar are the most notable examples of particular plants which, if not actually unknown in the western Mediterranean at earlier dates (e.g. the orange was present in Italy in Roman times and sugar may have been grown by them in Spain), certainly were not cultivated on a large scale until Islamic times.

From the Ebro right round to the Guadalquivir, Moslem engineers built river dams, usually to feed irrigation canals but also to meet the needs of power and water-supply as well. A conscious attempt to re-create in Valencia or Cordoba or Murcia the luxuriance of the Damascene Ghuta required the establishment of the same technical apparatus. Dams had to be built across rivers, primary and secondary canal systems were laid down with access to a suitable low-level drain, usually the parent or some other river, and a system of sluices, regulators and flow dividers had to be constructed. Actually, despite the attentions of later workers, the systems established in Spain are the best record we have of both the technology and the day-to-day operation of the typical Moslem irrigation system.

Extracting water from rivers required the construction of a diversion dam or a series of them. The size of the *huerta*, the régime of the river and the local topography, were all factors in determining whether to use one structure or a series. On the Rio Segura, a very large area around Murcia was watered from just one dam located upstream from the city. But lower down the same river, where it flows across the flat coastal plain between Orihuela and Guardamar, irrigation on a comparable scale was sustained by a network of low diversion dams. The same was true of Valencia whose basically Moslem irrigation system is thriving today and has in the past been a favourite topic of irrigation historians, not so much for its technology as for its interesting social and administrative features. Thomas Glick has recently used the Valencian example to counteract the Wittfogel thesis, although in this connection the scale of the engineering of Valencia's irrigation may be significant. Eight small dams and a set of short canals (the longest is 20 kilometres) of modest cross-section is a far cry from the massive works of Iraq, India or Ceylon.

The irrigation systems of rivers like the Segura and Turia are generally typical of the installations on the Mijares, the Jucar, the Guadalquivir and other rivers. When the south of Spain was ultimately reconquered by Christian kings, they inherited a land highly developed in irrigation networks. Naturally they were at pains to preserve not only the works but also the framework of their operation and maintenance. From the famous edict of James the Conqueror of 1239, issued just one year after he had overrun Valencia, the concluding lines are revealing: 'And so you may irrigate with them [the canals] and take waters without obligation, service or tribute; and you shall take these waters as was established of old and was customary in the times of the Saracens.'[3]

In the course of time medieval irrigators brought about developments of their own, organizational as much as technical, but unfortunately for the historian it is very difficult to judge exactly what these were, given the lack of Arabic records to describe the situation prior to the Reconquista. The dams of the Turia and the Segura, so firmly built into their foundations on beds of piles, so well shaped to withstand flood flows and erosion, and so carefully provided with special sluices to clear the canal headworks of silt, have Moslem origins, but equally their special features certainly contain an element of Christian evolution and innovation.

In another section of this book (Chapter 7) some space is given to Roman reservoirs impounding water for public supply, the most notable instances being the superb pair of earth dams at Mérida. It is interesting that this precedent in no way impressed itself on Spanish Moslem engineers. The requirements of irrigation, water-power and water-supply were always met with diversion dams of small size; everything was done according to the Middle Eastern pattern. Moslem society in Spain never experimented with shifting the basis of irrigation agriculture away from big rivers. The most radical innovation of Christian Spanish irrigators was just such a change.

Thomas Glick has demonstrated how smoothly the management of agriculture and the manipulation of irrigation passed from Moslems to Christians.[4] A key figure in the earliest phase of this transition was the 'Conqueror', James I of Aragon, and to him tradition ascribes the establishment of irrigation works on the River Ebro and its tributaries towards the end of the thirteenth century. Among these efforts is the first example in Spain (and in Europe for that matter) of a high dam built to impound irrigation water. At Almonacid de la Cuba at the head of a deep rocky ravine on the Rio Aguasvivas still stand the massive and ugly remains of the medieval dam and a later heightening. Both structurally and hydraulically crude, it is nevertheless a mighty monument to a new phase in Spanish irrigation history and a potent reminder of the ultimate curse of all reservoirs – silt. At Almonacid siltation is total; the reservoir now has no capacity to store any water at all and has long since been abandoned.

Almonacid de la Cuba was the first of a line of large Spanish dams. Quite rapidly the manifest structural and hydraulic defects of the pioneer were rectified and ultimately improved. The basic improvements are all evident in the famous Almansa dam, a structure built in 1384 and subsequently, in 1586, heightened and modified. Structurally it represents an important development because it makes use of the arch principle, the first dam in the West for which there is positive evidence of this innovation. The means to draw water from the dam was crude, a single low-level tunnel discharging through a bronze sliding gate, but provision was also made to de-silt the reservoir. Periodically a low level gallery, 1·3 metres wide and 1·5 metres high, was opened to its full extent by removing a set of closing beams, thus enabling a sudden and violent discharge of a very large quantity of reservoir water to flush away large amounts of silt, particularly that which had accumulated close to the dam. This technique has proved very effective. Although the Almansa reservoir is not entirely free from silt beds, the combination of a de-silting gallery, the sixteenth-century heightening of the dam wall, and a separate outlet tower built in 1921, allow the reservoir to function even today. A record of nearly six centuries of use is impressive by any standards.

Elsewhere the history of Spanish dam-building, with irrigation always the main inspiration, has been set out in more detail than is possible here.[5]

The huge Tibi dam (Plate 3), the highest in the world for nearly three

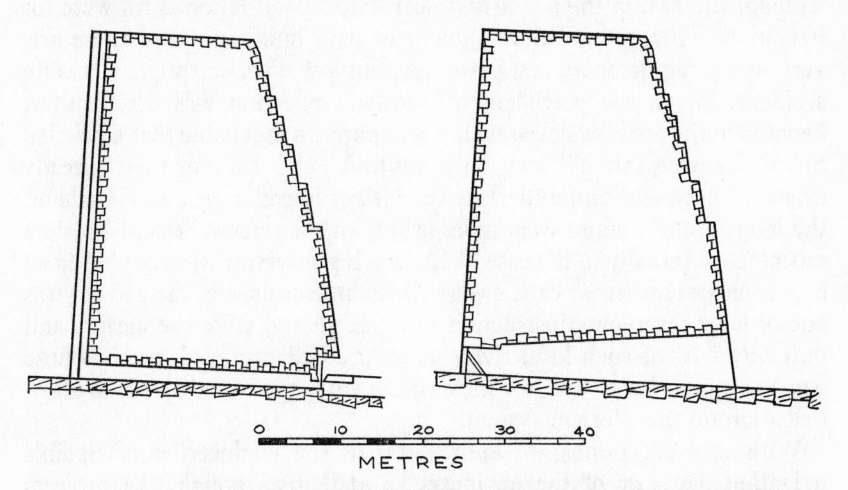

Figure 5 Two cross-sections of the Tibi dam. The left-hand drawing shows the vertical shaft, horizontal tunnel and sluice gate used to draw water; the right-hand drawing shows the tapered scouring gallery closed with wooden beams.

centuries, was completed in 1594 and supplied, as it still does, irrigation water to the land bordering the Rio Monegre between Tibi and Alicante. In the Tibi dam (see Figure 5), the hydraulic arrangements show a marked improvement over Almansa. By building a vertical outlet shaft over the full height (140 feet) of the dam, with multiple openings connecting with the reservoir, water could be drawn off no matter how much silt had accumulated. But periodically, as at Almansa, silt was flushed away through a large low-level tunnel. Hydraulically the Tibi arrangement is much improved because here the gallery is tapered so as to make use of the principle of the diffuser, the jet of escaping water and silt flowing that much faster and therefore with greater effect.

Although curved in plan, the Tibi dam is so thick as to depend hardly at all on arch action for its strength. By the seventeenth-century Spanish dam-builders were sufficiently confident to utilize arch action alone. The slender wall of the Elche dam sweeps boldly through an angle of 120° and structurally this dam is as notable as any in Spain. Hydraulically it has been rather less distinguished. Unlike Almansa and Tibi, Elche has no proper spillway and siltation is rapidly taking over despite the provision of de-silting apparatus. Much the same is true of another seventeenth-century structure, the Relleu dam on the Rio Amadorio, nowadays so silted up as to be of very little use even though structurally it is still sound.

The sixteenth and seventeenth centuries witnessed important developments in dam-building in Spain. Apart from the structures mentioned, there were other dams built in the province of Alicante, some notable ones in the west of Spain in Extremadura, several in the vicinity of Madrid and Toledo, and two in the north-east near Tudela and Huesca. All were for irrigation. This massive development of dam building – and these are, very often, big dams for water-storage, not low diversion weirs – was no accident. Given the precedent of Moslem irrigation with its manifest benefits to the areas so developed, it was perhaps inevitable that Christian Spanish society should not only continue and extend what already existed, but in addition undertake the task of introducing a new element, the irrigation of regions even more arid than the river valleys of Moslem settlement. Actually in the case of Elche a big reservoir was coupled to an irrigation system whose canals were Moslem. But usually the pattern was one of large reservoir construction to intercept and store the meagre and untimely flow of such local rivers as nature had provided, rivers whose day-to-day discharge was, for most of the year, quite inadequate to allow irrigation on the Moslem system.

With rare exceptions we know little of the engineering of Spain's irrigation dams or of the engineers. Apparently, though, the projects cited were all initiated, organized and financed by local communities and built and maintained by them. On more than one occasion the demands of the projects were too great, temporarily if not permanently, and examples of lengthy interruptions in construction or operational failures long delayed in their remedies can be found.

The localized and private sponsorship of sixteenth- and seventeenth-century Spanish irrigation affords some further evidence that civil engineering of this type is not necessarily associated with the influence of strong, central administration. It is interesting, though, that by 1700 the momentum had gone out of irrigation developments and it was not to pick up again until the end of the eighteenth century. In the reigns of Charles III and Charles IV, two very large irrigation reservoirs were constructed on the Rio Guadalentin above Lorca. Neither was ultimately successful. The Puentes reservoir served irrigation needs well enough for eleven years but came to a dramatic and disastrous end in 1802 when the dam failed. The Puentes dam was at the time by far the biggest in the world, a huge masonry gravity dam 50 metres high and 282 metres long.

But the builders committed a grievous error of judgement in allowing the dam to be built on a foundation of wooden piles, a most uncharacteristic move in comparison with the high standards of foundation work in earlier dams. When the reservoir was filled right up for the first time, the water-pressure blew the foundations clean out from under the dam, a substantial part of which then collapsed, releasing at least 35,000,000 cubic metres of water within one hour. The effect downstream on life, land and property was devastating.

The Puentes dam's eighteenth-century neighbour, the Valdeinfierno dam 14 kilometres up-river, still stands. But not since early in the nineteenth century has it been used. The dam's hydraulic arrangement proved to be inadequate, de-silting the reservoir damaged down-stream property, and the reservoir bed was so porous as seriously to limit its capacity to hold water.

Just as the Spanish acquired the basic arts of irrigation as the legacy of an alien occupation, so the diffusion and transmission of techniques progressed further, to the New World, by the same process.

Irrigation agriculture began in South America in the mountain regions of southern Peru and northern Bolivia. Subsequently, it appears as an indispensable feature of life in the coastal river valleys of Peru where early American societies flourished from the fifth to the fifteenth centuries. Irrigation was sustained both by diversion dams feeding canal networks, often tens of miles in total length, and also, in a few instances, by reservoirs of modest size. It is interesting to notice that these developments serve to emphasize how basic a technology irrigation is, and also how universal are the techniques employed. Even though place and period are different and contact entirely absent, man will arrive at the same solution to a given problem when it is posed in similar conditions. The huge irrigation networks built in Ceylon using massive tanks and canal networks are suggestive of the same thing.[6]

Of the two great pre-Columbian civilizations of Central America, the Mayas and the Aztecs, the former were irrigators only on the smallest scale although quite elaborate engineering was undertaken for water-supply. In Mexico the Aztecs were the first to struggle with hydraulic problems and develop irrigation. In and around their capital city, Tenochtitlan, on the western edge of Lake Texcoco, the ingenious *chinampas* system was evolved,[7] a highly successful irrigation technique which was, however, often threatened and occasionally severely disrupted by floods from the salty Lake Texcoco.

When the Spanish Conquistadors arrived in Mexico they found themselves in a land which was ripe for the introduction of their expertise in irrigation. Prominent in the construction of irrigation works was the influence of the Church, whose priests often took charge of construction. Many rivers were dammed in order to divert water to arid regions, notably the River Lerma, which not only fed the massive reservoir called the Lake of Yuriria but also sustained the irrigation canals of Salamanca, a town in central Mexico founded in 1603 by the Viceroy, the Marques de Montes Claros.

Logically, all the early Spanish efforts at irrigation in Mexico relied on diversion dams; large reservoir dams do not appear until the eighteenth century. Conceivably, it was the followers of Pizarro and Cortez who were particularly influential in introducing big dams. Both of the great conquistadors were natives of Extremadura, the part of Spain where the buttress dam was first built in large numbers, a notable example being the Albuera de Feria dam of 1747. It was precisely this form of construction which became characteristic of big irrigation dams in Mexico. Many of these structures still stand – a few in fact are still in use – and among them are a few of very baroque appearance, some of the most ornate and richly decorated dams ever made. At least one Mexican buttress dam repeats the Extremaduran concept of installing a power device within the body of the dam itself.

From Central America the Spaniards carried irrigation technology northwards into California, Arizona, New Mexico and Texas; and in this phase the influence of the Church continued to be paramount. A chain of missions established from Los Angeles to San Antonio prompted a good deal of dam building and irrigation, and here and there remnants, such as the Old Mission and El Molino dams on the San Diego river and the Espada dam on the San Antonio river, still survive.

The problem of irrigating their American colonies was one which Spanish engineers could manage. That they utilized the full repertoire of techniques already commonplace in the mother country is evident from the canal systems, the dams and their distinctive ancillaries such as desilting sluices, the use in many places of water-wheels including the noria, and the appearance in Mexico of the qanat. But in the one hydraulic problem which taxed them above all others, Spanish engineers found themselves seriously lacking in experience. Having established Mexico City, the old Aztec Tenochtitlan, as their capital, they had to grapple with the problem of flooding from the huge body of lakes to the east. In rapid succession, in 1533, 1580, 1604 and 1607, heavy floods had seriously disrupted life in the city and severely damaged agriculture around it. Following the inundation of 1607 an engineer called Enrico Martinez (said by some to be Dutch, but also referred to as 'a humble printer'!) attempted to drain the northern lakes with a 4½ mile-long tunnel. This was a failure, as was the later construction of a variety of dykes and embankments by the Dutch engineer, Adriaan Boot. In the 1620s Martinez failed once more with his scheme for a drainage tunnel but, having been jailed for his incompetence, was released in 1634 to try again. This time the work he initiated was destined to be of some success, but not in his own time. His original idea of a tunnel was finally put through as an open cut late in the eighteenth century and at last Mexico City was secure from all but the worst floods.

The association of at least one and possibly two Dutch engineers with projects so far afield as Mexico serves to reinforce an observation made elsewhere (Chapter 8) in relation to water-powered pumping installations – that by the seventeenth century Dutch expertise and experience in hydraulic technology was widely recognized and frequently consulted. In no field was this more the case than drainage and land reclamation.

3

The Great Reclamations

In a few parts of the north of Europe, in parallel with their efforts in Italy, Roman engineers carried out some flood control and drainage works. In the English Fens and in Holland embankments were built to contain the sea and unruly rivers, while here and there canals, sometimes quite large, served to drain low-lying ground but may, as in the case of the Carr and Fosse Dykes, have been just as important as navigations. When the Roman legions left northern Europe these hydraulic works lapsed rapidly and completely, and in the case of the Fens an inhospitable and deserted region was to remain isolated and hostile until the late Middle Ages, while reclamation was hardly attempted until much later.

The Frisians were the first people in north-western Europe to take on the struggle with the North Sea.[1] Dependent for centuries on their *terpen* – great mounds on which people congregated in time of flood – they progressed eventually to dyke construction and by the ninth century the first true polders were being formed. Further south than Frisia in Zeeland, Holland and Flanders a similar sequence was followed. Gradually throughout the Netherlands more and more dykes were constructed, polders were formed, and their drainage crudely achieved by sluices discharging at low tide. Progress was, typically, a function of the degree to which social groups could organize their labour and resources; and, interestingly, just as the church had been instrumental in prompting Mexican irrigation, so it was effective also in promoting Dutch reclamation. Some measure of how critical these efforts were to the Netherlands may be gathered from the fact that overall the battle was being lost. Although given communities in some places at certain times achieved a good deal in terms of flood protection and land reclamation, the sea was just as likely to undo in a matter of hours the achievements of men over decades. Late in the thirteenth century, for instance, the Zuider Zee was on balance getting larger, not smaller, and in 1287 floods occurred which were large enough to kill 50,000 people. But even if the problem remained formidable, the technology with which to solve it was evolving. According to whether the works were offensive or defensive in character and depending on the place, and therefore the materials available, various construction techniques found favour. Masses of boulder clay formed the basis of dyke construction and this was protected and reinforced with systems of osiers, reeds and seaweed. Subsequently, beginning in the fifteenth century, better protective methods were introduced using arrangements of piles leading ultimately to the *krebbingen*, a double row of stakes containing masses of faggots or fascines. Generally speaking, the use of stone pitching never became popular; although structurally very strong and durable, stone was far too expensive in the Netherlands for the scale on which dykes had to be built.

The earliest attempts at polder drainage remain rather obscure. Sluices discharging at low tide were certainly used and occasionally animal- or even man-power was used to lift water over the dykes, although by what means seems uncertain; apparently even such an ancient device as the Archimedean screw was not introduced into Dutch drainage until the late sixteenth century, and even then was only a limited success.

The classic Dutch water-lifting technique was the combination of windmill and scoop-wheel.[2] The origins of the scoop-wheel are obscure although the notion that animal-driven versions were utilized in the Middle Ages is certainly in accord with the basic simplicity of the device. In effect it is no more than an undershot water-wheel (or rather a very low breast-wheel) running in reverse, a water-wheel which is driven. In view of the generally widespread use of vertical water-wheels in medieval Europe, it is plausible that the process reversed should have found application in the Netherlands, although one should not underestimate the conceptual step required to visualize what to do. By contrast, one might reasonably propose an alternative line of development. If Spanish Moslem water-raising devices such as the shaduf and noria could prove capable of diffusion beyond Spain, then conceivably the Netherlands was one such area to be influenced. The noria as such could not of course be used because the flow of water required to drive one did not exist in the Low Countries. So Dutch drainers turned to wind-power.

Although so much has been written on the windmill, its European origins remain, in some respects, obscure. That the idea came from the East through the reports of Crusaders or via Spain or Sicily is no more than circumstantial. Eastern windmills pre-date any European device by 200 years, but whereas the former is a distinctive vertically-axled machine with direct drive to its mill-stones, the European windmill (first illustrated in *c.* 1270) featured vertical sails on a horizontal shaft driving through gears. If one postulates a westward diffusion of the *idea*, then a real difficulty is to explain how and why the machine itself underwent such radical re-arrangement.

The windmill as a corn grinder first appears in the Netherlands around 1200; the first evidence of wind-powered water-raising comes at the beginning of the fifteenth century and a type of windmill for the latter purpose developed from the former. The traditional medieval corn-grinding windmill was the so-called post-mill; a box-like structure carried the sails and the whole ensemble could be turned on its central wooden pillar to face the wind. The water-raising wind machine which evolved from the post-mill soon after 1400 was the famous *wipmolen*, the rotatable upper section being somewhat smaller, the fixed pyramidal base rather larger. The *wipmolen* featured two right-angled geared drives: one carried the drive from the horizontal wind-shaft to the central upright shaft, the other converted the latter's motion to a horizontal drive to the vertical scoop-wheel (see Figure 6).

By the nature of its design and mode of action a scoop wheel-cannot lift water more than about one quarter of its own diameter. Assuming a scoop wheel of some 8 metres diameter, at a maximum, then a lift of 2 metres was the best that could be achieved and usually it was up to 50% less. For many purposes, and particularly in the fifteenth and sixteenth centuries when drainage was just developing, lifts of up to 2 metres sufficed. If they did not, Dutch engineers resorted to the more elaborate and costly expedient of lifting in stages, one *wipmolen* raising water to a basin from which the next windmill took it a stage higher. Such a series of *wipmolen*, called a *molengang*, could comprise three or even four machines, and in this way water could be lifted as much as 20 feet to the level of the ring canal. From there it was drained into a river or directly into the sea.

The windmill achieved its ultimate development in the hands of Dutch mill-wrights. The biggest Dutch windmills, whether for pumping, sawing or fulling, were massive prime movers by any standards and a monument to the skill and know-how of their creators. But for all that they achieved, particularly in the critical task of polder drainage, one is bound to observe a basic drawback to wind-powered water-raising. Both the scoop-wheel and the Archimedean screw are very low-lift machines, evidenced in Holland by the fact that series of them were often needed to achieve lifts of a few metres, heights that with other elementary machines such as the noria or saquiyah would be regarded as nominal. Why were these alternatives not in fact used? Presumably the reason lies with the windmill itself. It is a very low power machine.[3] Such estimates as have been made seem to show that the biggest windmill operating under the most favourable conditions might develop around 50 h.p., of which about half was eventually delivered to the scoop-wheel or screw. (This would mean about 2,000 litres of water raised one metre every second.) It is in the nature of wind-power that it cannot be concentrated, so that in order to achieve

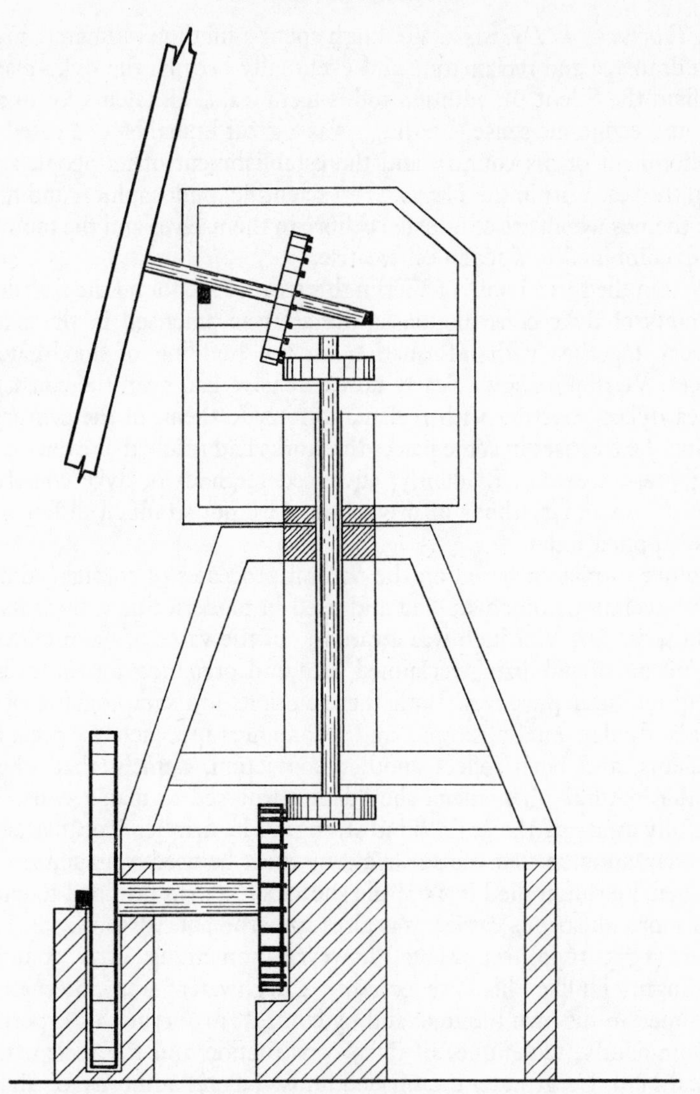

Figure 6 The essential layout of the wind-powered scoop-wheel in its early form.

higher performance it is the use of more than one windmill which is crucial, not the use of a single device of greater lift driven by a single windmill of greater performance.

Famous names are associated with the evolution of Dutch land reclamation and drainage technology. Of the personal life of Andries Vierlingh (1507–79) of Brabant we know very little but, fortuitously, a good deal is known of his work from the important manuscript which he prepared in the 1570s, but which was not published until 1920 when it was given the title *Tractaet van Dyckagie*. Vierlingh spent a lifetime intimately involved with drainage and reclamation and eventually became the dyke-master of William the Silent. In addition to his technical skill, talents for organization and economic sense, Vierlingh was a great humanist dedicated to the development of his country and the establishment of its people's safety from the sea. Within the *Tractaet* run scientific, philosophical and humanistic themes which are of unusual calibre in themselves and the more so for being combined in a technical treatise.

Within the three books of Vierlingh's treatise are found the first detailed accounts of dyke construction as the art was practised in the sixteenth century together with information on the building of flood-gates and sluices. Vierlingh shows clearly how extensive had been the construction of sea dykes, together with reclamation inside them, in the century preceding the treatise; in some places the works had reached or even exceeded their present extent. Evidently, advanced methods of dyke construction were in use and methods of protection were not so much different from those applied today.

Polder formation based on the natural processes of coastal sedimentation, a technique much studied and used in modern times, finds its place in the *Tractaet*. Vierlingh was aware too of the value of plant cultivation as a means of stabilizing reclaimed land and preparing for its future use for agricultural purposes. Both these notions are symptomatic of Vierlingh's fundamentally simple and economic approach to reclamation problems, and both reflect another conviction, namely that whenever possible natural phenomena should be harnessed to man's ends. While painfully aware of the sea's 'blind strength', he firmly advocated patience and adaptation so that the sea could in effect be used against itself.

When Vierlingh died in 1579 his successor, a man destined to enjoy an even more illustrious career, was already born. Jan Adriaanszoon Leeghwater (1575–1650) represents, however, something of a contrast to Vierlingh. Unlike his predecessor, Leeghwater was internationally acclaimed in his own lifetime, and by contrast to Vierlingh's expertise in, predominantly, techniques of dyke construction and the reclamation of coastal land, Leeghwater established himself as *the* authority on drainage by pumping. Born in North Holland at De Rijp, Leeghwater took up engineering as a young man, influenced probably by the problems of the lake-riddled region in which he grew up. Between 1607 and 1612 he laid the foundations of his future fame by successfully draining the huge Lake Beemster. Putting into practice his own conviction that 'the drainage of inland lakes was one of the most vital, most profitable and most sacred tasks which faced the whole of Holland', he proceeded over the next twenty years to pump dry most of the meres in North Holland.

Leeghwater seems not to have been an inventor of pumping equipment or a notable innovator in terms of techniques. Rather he brought together all the best features of both long and recently established practice, developed their best facets and above all established the means to drain land on a much more massive scale, the sort of scale that between 1615 and 1640 laid dry some 80 square miles of inland meres. For projects on this scale Leeghwater made extensive use of the *molengang*, building large windmills to drive, in the main, scoop-wheels. By far his most ambitious proposal for the exploitation of multi-stage drainage is to be found in his plans of 1629 to drain the Haarlemmermeer, 150 square kilometres in extent and about 5 metres deep. He planned to use no less than 160 windmills (in twenty groups of eight each) to lift water into a *ringvaart*, 50 kilometres long. The success of this project would have crowned Leeghwater's career in the biggest possible way but it was not to be. So mighty an effort, no doubt within the compass of Leeghwater and his colleagues, was too much for the pockets and courage of those in administrative and political power. In fact, not until 1852 was the Haarlemmermeer finally drained. Nevertheless, there is a sense in which Leeghwater is commemorated by his association with the Haarlemmermeer. In 1641 he published his famous *Haarlemmermeerboek* (it ran through a dozen editions right up to 1838) wherein one can gain an insight into this remarkable engineer, not perhaps possessing the breadth of philosophical and humanitarian interests of Vierlingh or the scientific accomplishments of another contemporary, Simon Stevin (1548–1620), but a realistic engineer of great talent and vision and at the time without equal.

Demonstrably the period just before and for half a century after 1600 was a high point in the evolution of land hydraulics, in Europe generally and especially in the Low Countries. To the names of Vierlingh, Stevin and Leeghwater can be added those of numerous other notable practitioners: J. B. Veeris, Dominicus van Melckenbeke, Symon Hulsbos, Nicolaas de Witt, Cornelis Meijer, Humphrey Bradley and Cornelis Vermuyden. Some remarks on these last two are now in order because with them comes the question of developments outside the Netherlands, for both Bradley and Vermuyden are best known for their projects in England and in France.

Although areas such as Sedgemoor and Hatfield Chase are not unimportant parts of the story of land reclamation in England, the focal point is the Fens.[4] The factors which eventually prompted attempts at Fen reclamation were not those which had prevailed in the Low Countries. Safety from the sea was far from being a crucial issue, while the formation of new agricultural areas was hardly a pressing need in a country of modest population already well provided with farm land. Down to medieval times, and excluding debatable projects of the Romans, only the Fenland abbeys had been instrumental in attempting Fenland drainage and their efforts were no more than haphazard, local and temporary. With the Dissolution of 1540, monastic influence was removed and came to be replaced by that of speculators and adventurers; and quite quickly it was the profit motive which stimulated a sustained and in the short term successful sequence of drainage schemes.

In 1589 Humphrey Bradley[5] of Bergen op Zoom submitted to Lord Burghley his proposals for draining the Great Level, later on to be known as the Bedford Level. Bradley's scheme reflected Dutch experience and know-how in two important respects which differentiate significantly his plans from all earlier ones. He knew full well that only a comprehensive drainage scheme for the *whole* region proffered any chance of success; and in order to draw up details of such a scheme, particularly in so very large a flat area, an accurate survey was essential. As a result of his survey one conclusion in particular emerged. Bradley saw that the lie of the land was such that given a proper set of drainage channels gravitational flow would be sufficient 'to redeem the land from the waters'.

Bradley's proposals were not taken up at the time. There were various obstacles to do with finance, disruption of traditional livelihoods, and a degree of antipathy towards Bradley himself. Not even Bradley's offer to use Dutch money, men and materials made any difference.

While Humphrey Bradley abandoned his English enterprises and went to France, the arguments over the Fens murmured on for forty years. In 1630 English speculators were at last ready to make a start and as engineer in charge of their operations they chose Cornelis Vermuyden, who had learned his hydraulic engineering in Zeeland and had already been involved in some English projects at Dagenham, Windsor Park and Hatfield Chase.[6] It is interesting to note here that in these early projects Vermuyden's efforts were not an unqualified success. This may seem odd in view of a previous claim about the excellence and experience of Dutch reclamation engineers. It must be born in mind, however, that conditions in England were not an exact replica of those in the Low Countries. Of course, Dutch engineers were head and shoulders above anyone else who could possibly have been employed, but nevertheless on the Great Level in the Fens they came up against some new facets of the land drainage problem which were beyond their experience.

Fundamentally, the Fen problem was more complex because of the existence of an elaborate system of rivers and unusual outfall conditions in the Wash (Figure 7). Embankment of polders and mere drainage was one thing; drainage of flat country traversed by four sluggish rivers – the Nene, the Ouse, the Welland and the Witham – was something rather different. And neither Vermuyden nor anyone else was aware of the peculiar soil conditions which are a feature of the region. Vermuyden's drainage scheme was based on the basic feature measured by Bradley and his surveyors, namely, that since no part of the Fens was then lower than 'high sea-level', gravitational flow along drainage channels would suffice. Judged purely in that light Vermuyden's scheme was a success. For many

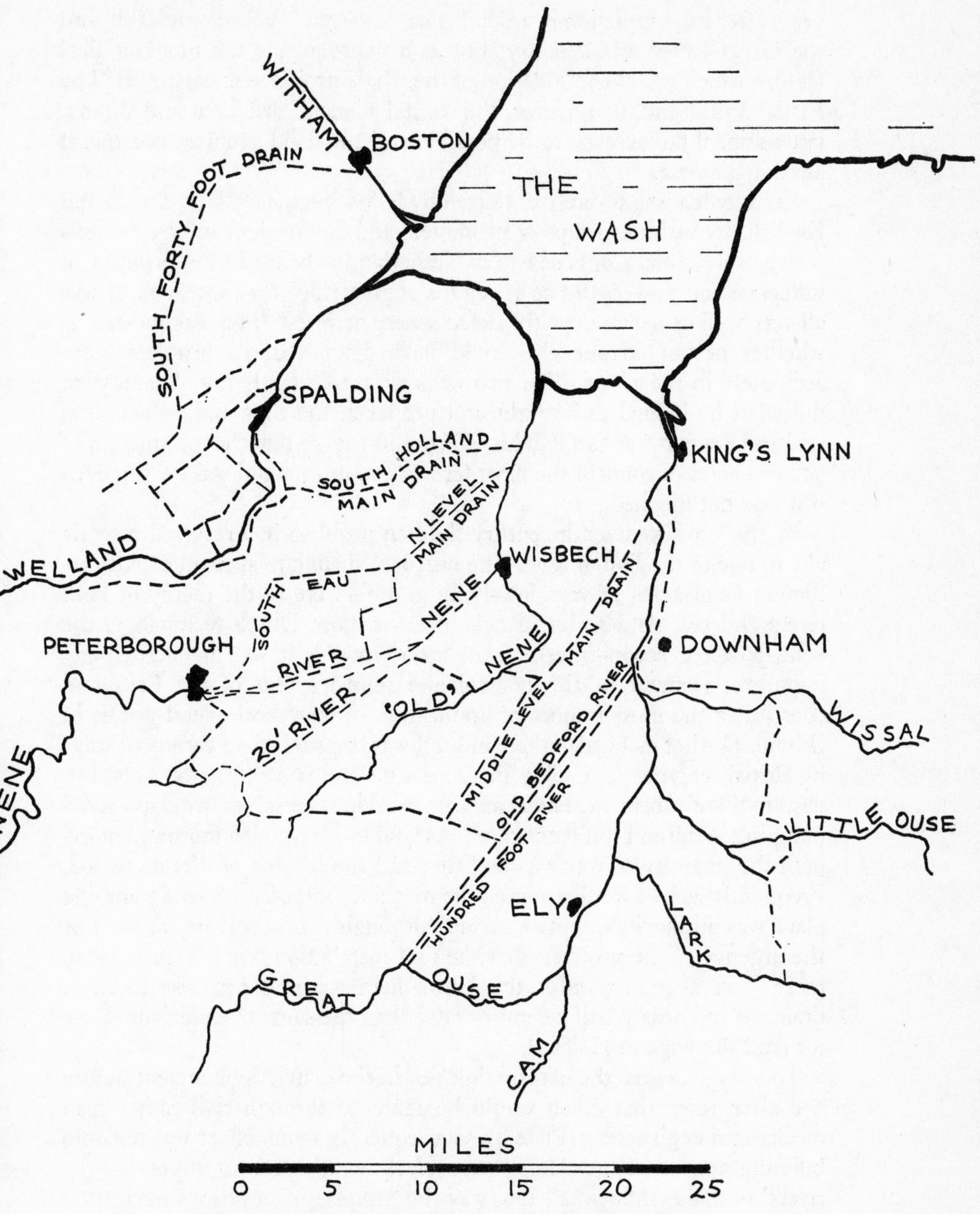

Figure 7 The system of rivers and drains in the Fens.

years after its completion in 1653, his canal system worked admirably and the Great Level was laid dry, but as a consequence the topographical feature which was the foundation of the whole project was destroyed. The Fenland peat and, to a degree, the coastal silt-land dried out and shrank, gravitational flow ceased to be possible and the flood problem became, if anything, worse.

Vermuyden was severely criticized by fellow engineers, both Dutch and English, by various groups of promoters and landowners and by factions which were simply opposed to a Netherlander being in his position of influence and power. But so far as his engineering was concerned, it was all very well to accuse once the defects were manifest. The real question is whether or not anyone else could have diagnosed the problem more accurately in the 1620s. The answer is almost certainly not. Vermuyden did what he judged to be right and proper at the time and, when all is said and done, it was through his efforts and energy that the reclamation of 700,000 acres of some of the most fertile land in England was begun, even if it was not finished.

By the late seventeenth century the Fen problem had reverted essentia ally to one of the Dutch type. The peat had shrunk to such an extent tht-thousands of acres of very low-lying ground were at the mercy of both rivers and sea. And so into England came more Dutch technology, the wind-powered scoop-wheel. From 1700 onwards it was an increasingly important element in the control and improvement of the Fens and constitutes the most important application of large-scale wind-power in Britain. During its heyday the windmill was improved in a variety of ways by British engineers, notably in the design and control of the sails, but despite these efforts, ingenious and practical in themselves, wind-powered pumping in the end could not cope.[7] As land levels receded more and more, both the quantity of water to be lifted and the heights of lifts increased. Frequently lack of wind rendered the machines impotent. Steam pumping plant was an inevitable introduction, although it is surprising in view of the urgency of the problem that the first installation was as late as 1820, more than a century after the Newcomen engine's first use in mine drainage and nearly half a century later than the same machine's first use for land drainage in Holland.

To a large degree the explanation lies in the conviction, current before and after 1800, that much would be achieved through civil rather than mechanical engineering (Plate 4). Consequently much effort was put into building new and larger 'cuts', particularly with a view to improving the rivers' outfalls. Influential, too, was the promise of improved navigation and consequently better access to the sea for Fenland produce, and to a degree this improvement was realized.

Initially, steam-powered pumping was applied in the traditional way, via scoop-wheels. But inherent defects in the scoop-wheel concept were never resolved by any change in design or construction. In particular, they were very wasteful of power, impossible to use for high lifts unless several were employed in series, and worst of all they were unable to accommodate changes in level at a given site, whether it was a fall in level on the intake side or a rise on the outlet side. In the nineteenth century an idea that had been proposed on paper at frequent intervals for nearly a century and a half was at last realized as a practical possibility. The earliest centrifugal pump for serious commercial use was the Massachusetts pump, first used in 1818 and illustrated in Figure 8. Centrifugal

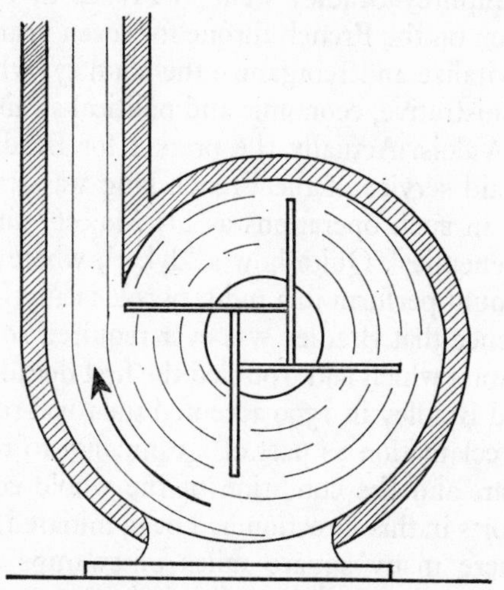

Figure 8 The Massachusetts pump of 1818. Water entered at the axle level of the four bladed impeller.

pumps have so many advantages – high speed, small size, the capacity to shift dirty and even muddy water, and good performance over a wide range of heads and quantities – that interest in their exploitation rapidly gained momentum.[8] Other factors which encouraged their development were the increasing availability of suitable motors – higher speed steam engines, gas engines and electric motors – and the evolution of the pressure turbine whose basic reversibility began to be examined by James Thomson around 1850 and was more formally expressed later on by Osborne Reynolds. The first centrifugal pump to be applied to Fen drainage was the design of J. G. Appold whose machine was successfully tested against others at the Great Exhibition of 1851. Once the advantages of the horizontal centrifugal pump had been demonstrated – on Whittlesea Mere in 1852 – the scoop-wheel's days were decidedly numbered although some diehards clung to the idea for a while. But most new pumping installations were equipped with steam-powered centrifugal pumps and at the beginning of this century a further modernization was the introduction of the diesel engine, affording even more efficiency, convenience and flexibility.

The above outline of the development of drainage engineering and land reclamation techniques in the Netherlands and England is indicative of the important role played by such work in the evolution of civil engineering, and to an extent mechanical engineering, in these countries. Was anything similar true for other countries? The cases of France and Italy will bear investigation.

Following his brief and in the final analysis unproductive interlude in the Fenland, Humphrey Bradley went to France in 1596. By this time Henry IV had been on the French throne for seven years and had already done much to revitalize and reorganize the country, which had been left in such dire administrative, economic and political straits with the demise of the House of Valois. Actually the pretext for Bradley's introduction into the highly paid service of the French king was strategic: to 'advise and instruct him in such operations as his Majesty should take for the damaging of his enemies'. Quite how a 'dyker', whatever his experience and reputation, could perform in a military role is anything but clear, nor is there any evidence that Bradley was ever required so to do. But in any case the internal wars which had troubled the first decade of Henry's reign soon subsided and Bradley in 1599 accepted the royal request to set about large-scale land reclamation as part of a campaign to rehabilitate certain areas in particular, and the condition of the whole country in general. Already some efforts in this direction had been initiated, around Bordeaux for instance, where many square miles of swamps and marshes had acquired an evil reputation for ague and malaria, some epidemics leading to an appalling loss of life. But prior to the seventeenth century such projects as were conceived and occasionally begun, at Bordeaux and other places, came to nothing; they failed due to lack of technique and because the unhealthy conditions were quite capable of slaying the work force.

Just as English conditions were unlike those prevailing in the Netherlands, so France was different again. Particularly in the south and southwest where higher temperatures and humidities occur, the question of disease was notably worse. As in England, but unlike Holland, the technical problem was fundamentally one of draining low-lying land in the vicinity of slow-moving rivers, but with complications. The reclamation question in England was confined, essentially, to one self-contained area, the Fens, but in France there were numerous such regions, not as large individually but in total considerably larger. And many were inland regions rather than coastal or estuarine.

The work of Bradley and his fellow Dutchmen who joined or succeeded him seems not to have utilized anything that was technically novel or advanced. But the organization and administration of the schemes undertaken was a radical improvement on contemporary English efforts. In contrast to the speculative and private nature of the English ventures, France treated her reclamation problem in the correct way, as public works, backed and financed by the government. This properly co-ordinated and officially sponsored approach was the underlying reason for the considerable progress achieved in French land reclamation and relief from the *peste* in the seventeenth century. It helps to explain, too, Bradley's successes in France as against his failure in England.

Bradley's compatriots who succeeded him in French reclamation work seem to have been active for a generation or two and then the influence of the Netherlands declines rapidly. But French engineers were well able to take up the challenge and in the eighteenth century there were attempts, for the first time in France, to reclaim land from the sea.

The work of Dutch land reclaimers and drainers in Italy cannot by and large be judged a great success or of much importance. Various factors combine to account for this state of affairs. In the first place the problems of Italian land hydraulics, unlike those in the Netherlands, England and France, did not attract serious attention for the first time in the sixteenth and seventeenth centuries. The problem was a traditional one of much longer standing which as early as Etruscan and Roman times had stimulated a variety of projects, some of which were notably successful. Typically, Roman works were uncared for in the Dark Ages and they suffered accordingly. But the Middle Ages witnessed a revival, usually at the instigation of monastic communities, sometimes by powerful city states.[9] Thus in Piedmont and Lombardy several attempts to drain key areas in the Po valley were in the hands of either Benedictine or Cistercian houses while in and around Milan the power and status of the Lombard League was responsible for irrigation dams and canals. Beginning in the twelfth century with works on the River Ticino, these irrigation schemes gradually extended to other rivers which drain into the Po from the northern lakes.

In central Italy monastic influence was active but ineffectual at an early date before the challenge was taken up by the Papal State. A succession of Popes embarked on reclamation projects in both the Roman Campagna and the Pontine Marshes during the sixteenth and seventeenth centuries, culminating in the mighty efforts of Sixtus V between 1585 and 1590. A large area between Rome and Terracina was successfully laid dry and made healthy by a system of drains and main canals. But neglect and local antagonism to papal influence combined to promote rapid deterioration of the best work thus far achieved. At Ravenna, Ferrara and Aquileia the late sixteenth century also witnessed land reclamation, but here too a decline was almost immediate. Another contributor to Italy's long experience of hydraulic problems was Venice. Quite apart from the unique hydraulic situation of the city itself, the fifteenth and sixteenth centuries saw efforts to control, drain and reclaim the mouths of the rivers Piave, Brenta and Adige.

Another factor militating against the effectiveness of Dutch engineers in Italy was the fundamentally unfamiliar nature of the problem. The Alps and the Apennines are the sources of many rivers, snow-fed torrents in their upper reaches but changing into sluggish streams, often wide and meandering, as they cross coastal plains to reach the sea. It is the régime of its rivers which has been and still is the determining factor in the preoccupation with, and the development of, hydro-technology and hydraulic science in Italy. The fundamental requirement has always been river training and regulation. On this has depended the provision of irrigation, the prevention of flooding, and the drainage of low-lying ground. The Po, the Arno and the Adige, to name three of Italy's most unruly rivers, have been the objects of a developing river control technology for as long as there have been social groups whose health and agriculture were threatened. From the Middle Ages onwards, control of the Po and its many tributaries, especially those in the north, was based on systems of dams and canals and protection and control works made up of pile-dykes, masonry walls, stone pitching and systems of fascines and wooden mattresses. It can be said that gradually the Po was brought under control except in the most exceptional circumstances, and these remain potentially destructive even today. The year 1967 bears witness that the same thing is true of the Arno, a river with a long and notorious record of flooding, particularly at Florence.[10]

The notably well documented history of the Arno's floods shows how often the Florentines have had to reconstruct their bridges and rebuild the river's banks. In 1495 Leonardo da Vinci was the first to propose a comprehensive plan for the river throughout its length, embracing such concepts as flood control dams up-river and canalization lower-down. But these schemes and similar later ones in the sixteenth century were not prosecuted.

Clearly, then, it would have been extremely difficult for Dutch engineers to work effectively in a country where there was already a body of experience in dealing with hydraulic problems which were of a basically different nature. In fact, of the few Dutchmen who made any sort of impact, it is significant that all were involved with the one Italian problem which their experience in northern Europe had prepared them to deal with: the Pontine marshes. In the seventeenth century a succession of proposals culminated in the promising scheme of Cornelis Meijer. The efficacy of Meijer's plans was never tested. Parochial opposition hindered everything he tried to do and despite being a Catholic he received little support from Popes Innocent XI and XII, who were nominally in favour of the work. Following Meijer's death in 1701, his son tried to continue the project but local opposition, disposed at times to wreck the very works which were intended for their benefit, thwarted him. The one notable legacy of Cornelis Meijer's labours is his memoir of 1678, *Del modo di secare le Palude Pontine*. It is in fact just one of many tracts on hydro-technology written in Italy in this period, the beginnings of a significant body of literature on the subject.[11]

Among a variety of fifteenth-century technical manuscripts, references to land hydraulics are rare until one reaches the wide-ranging researches of Leonardo da Vinci. As noted in Chapter 11, water held a special fascination for Leonardo and there is clear evidence that a book on the subject was very much in his mind. And consistent with our earlier remarks concerning the dominant role of rivers in Italian technology, one finds Leonardo preoccupied with the behaviour and control of rivers both in general terms and in a number of specific instances. His general studies range over water flow in channels of different forms, the control of rivers, the formation of shoals and the effects of scour, damage to river banks, and techniques of repair and maintenance. The topic had never before been subjected to such minute and close scrutiny and throughout are carefully drawn, even artistic, illustrations of the movement of water. However, it is not always clear whether the pictures are generalizations or depictions of specific instances met with in the course of some project or other. Typically, Leonardo failed, with one notable exception, to reduce his mass of detail to any general hydraulic law or laws. As Gille puts it: 'He was dealing with instances, not with laws.'[12]

Within the history of Renaissance engineering, Leonardo da Vinci is a mighty figure about whom much has been written, especially in the last few years. It is interesting to note, therefore, that opinions remain very divided as to his influence, the extent to which he was actively involved in practical engineering, and the role of his writings (and remember none was ever printed) in shaping the thoughts and work of his successors.[13] Quite the opposite is true of Antonio Lupicini (*c.* 1530–98), an engineer who practised widely in Italy, wrote six books and became famous throughout Europe. Lupicini's *Discourse on the Defences of the Po and Other Rivers with Artificial Embankments of Earth*, published in 1587 in Florence, comprises an analysis and description of river engineering for navigation and flood control. A later book is more specific. *Discourse on Flood-Protection at Florence* (1591) is devoted wholly to the vagaries of the Arno and the continuing failure of the Florentine *Ufficiali di Fiumi* to procure any measure of control despite numerous reports and the diagnoses of many engineers. Lupicini recognized three phenomena as the cause of flooding on the Arno: bank erosion, neglect of levees and rats. He proposed specific solutions to all three problems, notably the use of a circular assemblage of timbers and branches laid on the river bank and unrolled into the water to prevent the erosion of the Arno's banks, the river's most serious problem. Lupicini's recommendations in this respect are the earliest reference to 'mattress protection', a technique widely used in modern times.

Lupicini was an expert and a specialist. His expertise is reflected in his reports, the most authoritative and detailed yet compiled on river hydraulics. His specialism is apparent from the uniquely 'Italian' theme in his writings; everything is directed specifically to the problems of rapid and turbulent rivers, the beginnings of a technical literature quite different from Dutch writing on questions of coastal protection and land reclamation.[14]

The famous and fascinating *Le Diverse et Artificiose Machine* by Agostino Ramelli published in 1588 is contemporary with the works of

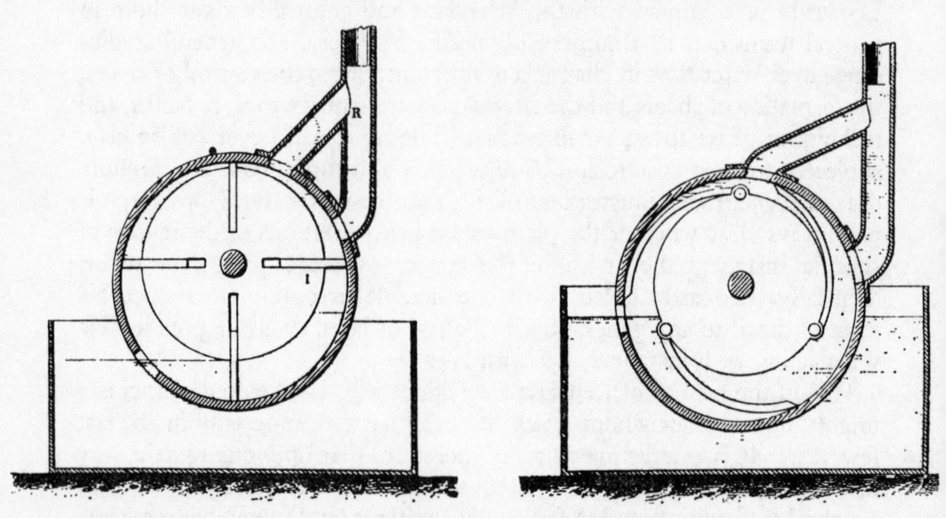

Figure 9 Ramelli's ideas for rotary pumps were still of interest in the nineteenth century. These two drawings are from Arthur Morin's *Machines à l'Élévation des Eaux* of 1863.

Lupicini, but it is a very different book. Whereas Lupicini's were serious textbooks and practical manuals in the style of the works of Agricola or Biringuccio, Ramelli's is a speculative and imaginative work wherein all manner of themes and systems are explored with rich variety and great ingenuity. Some ideas are sensible, much is variation for its own sake and rarely is any conceivable permutation not recorded in nearly 200 beautifully detailed drawings. Hydraulic engineering is a dominant theme and questions of land hydraulics are featured in fourteen of Ramelli's plates.[15] All of them deal with pumping devices, man- or water-powered, to drain low-lying ground. In addition to traditional systems of piston-pumps, chains-of-pots, norias and Archimedean screws, quite the most novel and interesting ideas are a sequence of rotary pumps of the displacement type (Figure 9). Within fixed cylindrical casings Ramelli shows eccentrically mounted rotors which draw water from a sump into a large space (between rotor and casing) and as the rotor spins the water is compressed by means of flat sliding vanes or curved hinged plates into a smaller space from whence it is discharged at a higher level.

In the context of land drainage Ramelli's exploration of the idea of rotary pumps is notable. Water to be pumped from swamps, marshes or flooded fields is liable to be charged with quantities of mud, sand, gravel, vegetation and even larger debris. In such conditions rotary pumps with their large flow passages and absence of valves have manifest advantages. However, it was not until the nineteenth century, with the advent of metal construction and suitably powerful, high-speed engines, that rotary pumps became a practical proposition.

In 1598 yet another Italian river was the subject of a flood study, this time the Tiber. The work was carried out by the architect Giovanni Fontana da Meli who succeeded in showing, to his own satisfaction at least, that the channel of the main stream was too small to carry the flood flow from the whole catchment area. In fact, da Meli had no clearer idea how to measure discharge than had Frontinus (see Chapter 7), but the basis of his approach and the general conclusions reached so impressed Benedetto Castelli that in 1640 he recommended that da Meli's report be reprinted.

On the basis of his classic work of 1628, *Della misura dell'acque correnti*, Benedetto Castelli has been acclaimed as the father of the Italian school of hydraulics and undeniably he is important for bringing together and publishing the first body of general theory on river hydraulics. In particular, he clearly set out the principle of continuity – 'Sections of the same river discharge equal quantities of water in equal times, even if the sections themselves are unequal' is his succinct statement – and the concept was soon absorbed into practical engineering.

Three important Italian treatises on land hydraulics followed that of Castelli; and all three were influenced by him.[16] Domenico Guglielmini's *Della Natura de Fiumi* of 1697, taking the now well accepted continuity principle of Castelli as a basis, proceeds to examine other features of open channel flow: the distribution of velocity in a stream, the flow conditions at a sluice or the inlet of a sloping channel, resistance to flow, and the nature of scour and bed erosion. Guglielmini failed to produce a complete solution to any of these problems but his conviction that such issues would yield to mathematical analysis is significant.

Giovanni Battista Barattieri's *Architettura d'acque* of 1699 deals principally with the problems of river regulation and perpetuates the erroneous view, although Castelli had been suspicious of it, that the distribution of velocity over the depth of a stream is parabolic with a zero component at the surface. It is strange that this manifestly absurd notion should have retained its adherents for so long.[17] It is repeated, for instance, in another Italian treatise of note, namely Paolo Frisi's *Del Modo di regolare i fiumi e i torrenti* of 1762. Frisi was a figure of some influence in eighteenth-century Italian scientific and mathematical circles and, while his contribution to river technology is nowhere very original, his compilation was highly influential in disseminating Italian knowledge and expertise.

Just prior to the close of the eighteenth century one highly ingenious

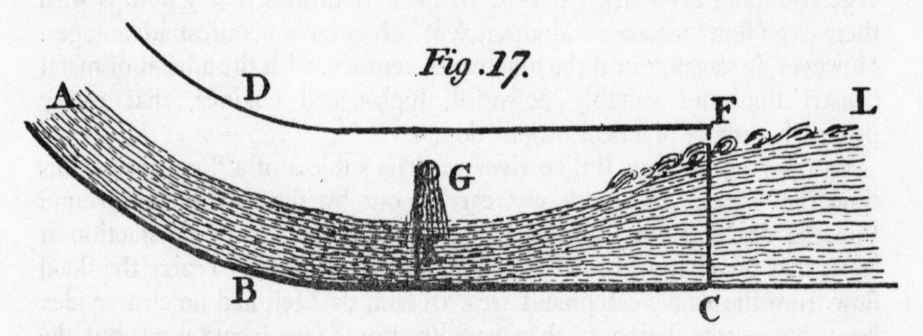

Figure 10 This diagram, Figure 17 from Venturi's *Recherches Expérimentales* of 1797, shows the famous Italian hydraulician's idea for utilizing the hydraulic jump to effect land drainage.

idea in the context of land drainage is worth noting. Among a wide variety of problems examined in *Recherches expérimentales sur le principe de la communication latérale du mouvement dans les fluides*, published in Paris in 1797, Giovanni Battista Venturi included the hydraulic jump. He was by no means able to explain this fascinating hydraulic phenomenon but did suggest how it might be utilized to effect drainage in the fashion shown in his drawing (Figure 10). Venturi *claimed* that he had successfully done this in a practical case.

4

Colonial Irrigation

By 1800 a very considerable body of expertise in land hydraulics had been evolved, tried and tested. Much remained to be theoretically analysed but the technology with which to achieve results was established and understood. Dutch engineers had contributed to matters of land drainage and reclamation through their work in the Netherlands, England, France and Italy. Italian engineers, by tackling the characteristic problem of their own country, had shown how a variety of problems to do with river regulation and flooding might be managed. Nor should we forget that in Spain, and to a degree in France and Italy, irrigation also was a highly developed art. Scientific explanations of techniques which were practically successful were generally lacking although the eighteenth century had produced many notable advances in the theory of hydraulics, with a whole variety of problems being isolated and solutions formulated. Few of these were wholly accurate or complete (and many were thoroughly defective) but in conjunction with the results of experimental analyses, to which eighteenth-century investigators had devoted considerable attention, hydro-mechanics had accumulated a substantial literature of value to both theorists and practising engineers.[1]

For the practical hydraulic engineer access to details of techniques and methods was greatly facilitated through this literature, not only in the original but more interestingly, in some ways, via their translations. Indeed it is very revealing how European-wide was the interest in hydraulics compared with that in, say, structural mechanics. Important tracts on arches or beam bending were sometimes hardly known even in their country of origin, whereas writings on hydraulics were widely disseminated even across the English Channel. By 1826 there had been English translations of the works of Castelli (1661), Mariotte (1718), Frisi (1818), Du Buat (1822), Venturi (1826) and Eytelwein (1826); famous tomes such as Belidor's *Architecture Hydraulique*, though not translated, were well known. In one respect the translation of Paolo Frisi is particularly interesting, and the point will take us forward into a new and important phase in the history of land hydraulics.

Frisi's *Del Modo di Regolare i fiumi e i torrenti*, first published in 1762, was rendered into English in 1818 by Major-General John Garstin. This was no mere act of curiosity; it was highly official. As noted earlier, the Indian sub-continent for a couple of thousand years has had an impressive history of irrigation based on rivers and dams, reservoirs and canals. Typically for India, details of the earliest developments are almost impossible to unravel but later on rather more is known, from the sixteenth century right down to the early years of the nineteenth century when British rule first became actively interested in irrigation, drainage and land reclamation. It was the British government who authorized Garstin's translation of Frisi, and paid for it. Lacking experience in the difficult techniques of large-scale land hydraulics, British engineers had to turn to whatever bodies of knowledge offered the advice and instruction needed to develop a vast and vastly underdeveloped acquisition such as India. The pattern was to be repeated throughout the nineteenth century and extended to colonial powers other than Britain and to one, the United States, which was not a colonial power.

On behalf of the Société Royale d'Agriculture of Paris, F. J. Jaubert de Passa made an extensive irrigation tour of southern Spain between 1816 and 1819 and reported his findings in 1823.[2] In 1850 Captain R. Baird-Smith visited Italy to 'examine the classic land of irrigation', a mission which was 'undertaken under instruction from the Honourable the Court of Directors of the East India Company'.[3] The 1860s produced a very distinguished set of travellers. The French engineer Maurice Aymard visited southern Spain with a view to developing irrigation in Algeria; Sir Clements Markham's observations on Spanish irrigation were printed by order of the Secretary of State for India;[4] between 1867 and 1868 C. C. Scott-Moncrieff carried out a 'tour of inspection of the irrigation works of France, Spain and Italy for the Government of India';[5] and J. P. Roberts' book of 1867 was likewise stimulated by Indian problems.[6]

In 1882 Claude Vincent published his report on irrigation and masonry dams, the result of yet another tour of Italy, France and Spain, this time at the instigation of the government of Madras.[7] William H. Hall, who was state engineer of California, visited Valencia, Granada and Murcia in the 1870s,[8] just prior to the extensive irrigation developments which began in the western United States in the last twenty years of the nineteenth century. When Germany embarked on grandiose colonial plans the irrigation issue figured in its thinking too. A report on Spanish irrigation prepared in 1888 for the German government was finally published in 1896.[9] Evidently, then, in the nineteenth century irrigation engineering was a prominent European technical 'export' to developing countries such as the United States and to colonial possessions such as India, Iraq, Egypt and Algeria. The situation is not without its irony too. More often than not, this new era of irrigation signalled a rebirth or revitalization of a basic technology which had already been expertly practised by the indigenous peoples (e.g. in India and Egypt) or by former occupiers (e.g. the Spanish in America or the Romans in North Africa).

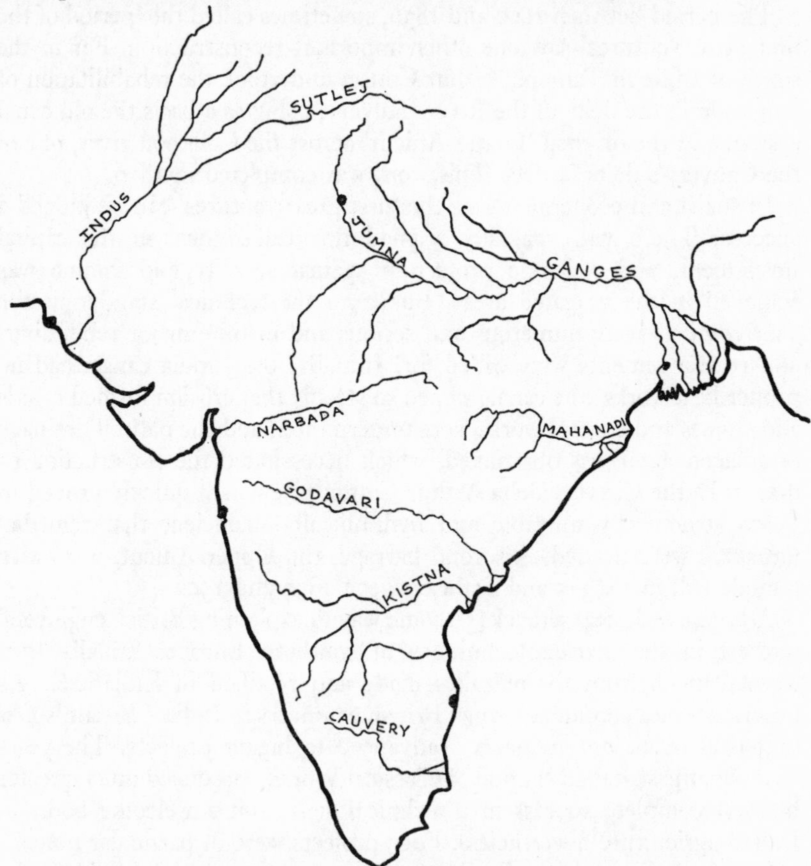

Figure 11 India and its river systems.

In British India the process was quite literally rehabilitation in a few of the earliest instances. In northern India several neglected and abandoned irrigation canals of Mogul origin were judged to be suitable for reconstruction. Prominent among these were the Jumna canals, lying to the east and west of the Jumna river (Figure 11). Begun in the fourteenth century and greatly extended and developed in the sixteenth century by the emperor Akbar, the Eastern and Western Jumna systems functioned efficiently until the early eighteenth century when they fell quickly into decay and ruins, the typical result of a collapse of administrative authority and the disorganization of the social order. By 1750 irrigation was defunct and not until 1810 did British engineers begin to examine its future. Serious work did not begin until 1817 and considering the unfavourable conditions under which the work was undertaken and the difficult engineering problems involved, it was no mean feat to realize full operation within not much more than a decade.

The period between 1800 and 1836, sometimes called the 'period of the first great ventures' saw one other important reconstruction. Far in the south of India in Tanjore, Arthur Cotton undertook the rehabilitation of irrigation in the delta of the River Cauvery, using as a basis the old canal system and the original 'Grand Anicut' across the Coleroon river, one of the Cauvery's delta outlets. This work was completed in 1836.

In social and economic terms the 'first great ventures' can be judged a success. There was generally a good financial return on the capital investment, and welcome protection against scarcity and famine was achieved in the irrigated areas. But from the technical standpoint the shortcomings were numerous and serious and in time major rebuildings and rearrangements were called for. Initially, the Jumna canals had no proper headworks; the canals sloped so steeply that erosion formed rapids and sluices and control works were undermined; and the natural drainage of adjacent land was obstructed, which necessitated the construction of drains. In the Cauvery delta Arthur Cotton's new dam quickly proved to be so structurally unstable and hydraulically inefficient that remedial measures were needed. A second barrage, the Upper Anicut, was twice remodelled, in 1843–5 and again between 1899 and 1902.

All these technical setbacks go some way to explaining British engineers' interest in the proven techniques of southern Europe. Equally they learned much from the mistakes made and rectified in India itself. As experience and confidence grew, British engineers in India – Britain's *first* school of irrigation engineers – advanced to bigger projects. The years 1836–66, the so-called 'Period of Classical Works', produced much greater but not complete success in a technical sense but a welcome boost to Indian agriculture nevertheless. Four projects were of particular note.

Between 1832 and 1833 India experienced a severe famine followed by several years of extreme scarcity. It was in an effort to prevent a repetition of the terrible conditions of the 1830s that Arthur Cotton, following his work on the Cauvery, was engaged on two other delta projects. Between 1844 and 1855 he planned and supervised the construction of extensive irrigation systems in the deltas of the Rivers Godavari and Kistna. Probably these were the best works of irrigation yet built in India, distinguished by the provision of well designed and permanent headworks and regulators, whose importance Cotton was now in a position to appreciate.

In north-eastern India Sir Proby Cautley took eighteen years to build the massive Ganges canal, all but 900 miles long and able to carry 6,750 cubic feet of water per second. Quite apart from the stupendous volume of excavation involved, the Ganges canal featured numerous dams and regulators, bridges, the huge Solani aqueduct bridge and locks; this aqueduct was so large that it was able to accommodate a useful degree of inland navigation as well.

Another work, the Bari Doab canal, was begun in 1857, a year or so after Britain's final annexation of the Punjab. The principal aim of the project was the irrigation of the arid central Punjab in an effort to ease the resettlement of large numbers of disbanded Sikh soldiers. Earlier lessons seem not to have been well learned in this case. When the canal was opened in 1859 it had no proper headworks and it took fifteen years of effort and setbacks to install them satisfactorily. Nor was the canal well graded – it was too steep as originally designed – and this too required attention at a later date.

Up to 1860 Indian irrigation was based on run-of-river concepts, occasionally featuring substantial permanent diversion dams (the Cotton technique) but more frequently using flimsy and temporary wooden weirs, a fatal weakness. In the 1860s Indian irrigation began to contemplate a new idea in relation to which the visits to Spain by Markham, Scott-Moncrieff, Roberts and Vincent are significant. Spain already possessed a centuries-old irrigation tradition based on large reservoirs featuring by far the most substantial number of big dams in the world. Structures such as Tibi and Elche figure prominently in the reports of Markham and Scott-Moncrieff (as they do in the writings of the Frenchman Maurice Aymard and the American Edward Wegmann), and sure enough at the end of the decade in question we find Indian irrigation adopting this markedly Spanish concept. It was a logical and necessary – and expensive – development, prompted by the need to install more seasonal and yearly flow regulation as an insurance against famine and protection against floods as well.

The most important of these early irrigation dams was the one across the River Periyar in Madras. A decade in the making, it was finished in 1898 and was the first British dam of such size – 173 feet high – to be made of concrete. It was in order to confirm the Periyar dam's design that Professor W. J. M. Rankine conducted his famous research into the stress analysis of gravity dams, which more than any other contribution lifted dam design from the realms of empiricism to rationalism.[10]

In addition to the construction of big reservoirs, irrigation in India in the last third of the nineteenth century was characterized by two other factors: colonization and protection. Certain areas of India were so barren and inhospitable that their inhabitation was, and always had been, virtually impossible. A specific aim of government policy was the opening up of arid regions in West Punjab and Sind by providing irrigation (and roads and railways). Here and there the resulting improvements were dramatic and the return on the capital invested extremely high. 'Protection irrigation' by contrast adopted a different attitude. Virtually regardless of initial cost and subsequent profit, some irrigation works – the Betwa scheme in Uttar Pradesh, for example – were installed purely as an insurance against famine, an issue to which attention had been drawn more powerfully than ever before by the great famine of 1877–8, which killed some four million people, a fearsome toll even by Indian standards.

There can be little doubt that British civil engineering touched a peak in India in the second half of the nineteenth century, a peak which is comparable to the one reached in Britain itself in the first part of the century. But whereas the earlier achievements have been written about at length, the Indian story remains virtually unexamined, a defect which we have no space to rectify here. Through irrigation and water-supply works, not to mention road and railway construction, it is clear that many benefits were brought to India's people. All the same, it should not be overlooked that certain wholly British aims were important too. Communications were established with military requirements very much in mind, although in time some such routes ceased to be of any real importance. There was a good deal of self-interested financial investment too, and this was true of irrigation works in the early years. The often considerable expense of their construction was recovered from the sale of water to the peasants and the returns were frequently very lucrative. But private enterprise eventually failed and after 1866 state financing took over.

India bred a line of irrigation engineers with international reputations. Some, such as Sir Arthur Cotton, Sir Proby Cautley and Sir Colin Scott-Moncrieff, remain best known for their efforts in India but others made their mark elsewhere. By far the most influential of these was Sir William Willcocks, a brilliant engineer of colourful and at times eccentric character, whose busy and eventful life could easily fill very many times the space we can spare him here.

William Willcocks was born in a tent in India in 1852; he was a brilliant student at the Thomason Civil Engineering College at Roorkee, and as a young man he worked for eleven years in the Indian irrigation service. His move to Egypt in 1883 was the first time he had ever left India. In Egypt Willcocks joined the staff of Sir Colin Scott-Moncrieff, also recently arrived from India to take charge of Egypt's Irrigation Service, which was in a manifestly chaotic condition. At the beginning of the nineteenth century, Mohammed Ali Pasha, in attempting to stimulate some sort of economic development in Egypt, became convinced that cotton promised to be a valuable export. He also discovered that its cultivation could not be supported by the age-old basin irrigation system: perennial irrigation based on storage of the Nile's annual flood was imperative.

Prior to the arrival of British engineers, attempts to provide perennial irrigation were very mediocre indeed. This is not altogether surprising. Initially, the construction of new and deeper canals and wells was successful up to a point. Beyond that, the damming of a river the size of the Nile was a formidable and unprecedented undertaking. Very naturally it was the Nile's two delta branches rather than the much bigger main stream which were tackled first, a procedure which even before 1800 Napoleon had seen to be desirable if the Nile delta was to be properly developed and the river fully controlled.

The first construction work began in 1833. But Mohammed Ali Pasha's engineers found themselves confronted with much more than they could handle; the project proved to be hopelessly expensive, and then a plague wiped out the bulk of the labour force. The scheme was finally abandoned. Fortunately, the Pyramids were saved; a ludicrous plan for dismantling them to furnish building stone was never begun.

French engineers next took up the challenge. In 1842 Linant de Bellefond and Mougel prepared plans for the Damietta and Rosetta dams and these structures were completed by 1861. They were good designs in principle but shoddily built and not very effective in operation. In particular, the Rosetta dam leaked badly in places and in 1867 a portion of it slid noticeably downstream. British experience in India was summoned to take charge of the situation and the Delta dams were made safe and effective.

By 1890 it had been decided to dam the Nile's main stream.[11] Much more storage capacity, commanding a larger irrigation area, was required in northern Egypt than could be provided from within the Delta. It was William Willcocks who was given charge of the survey of the Nile valley to seek out the best site and to recommend the form of dam required. The extraordinary zeal with which he undertook this task (and later on he was to repeat the performance in Iraq) is one of the most intriguing features of his approach to the work. He spent three years examining 800 miles of the Nile valley in company with his team of Egyptian and British surveyors and engineers. He seems rarely to have left the desert and developed a sense of fanatical devotion to his mission. Nothing was allowed to deflect him from his purpose. He eventually reached the point where he dispensed even with a tent at night and his diet came to consist of the appalling combination of rice, apricots, whisky and water. His days began with lines memorized from the Bible, John Bunyan or Shakespeare; they ended with the writing of technical reports, long, detailed and careful analyses of every aspect of the Nile's régime and topography.[12] The result is well known. In 1894 Willcocks recommended that a Nile dam be built at Aswan (Figure 12), a long, straight gravity structure of conventional masonry construction founded on granite. Unconventional, though, was the dam's hydraulic design. It was to be fitted throughout its length with

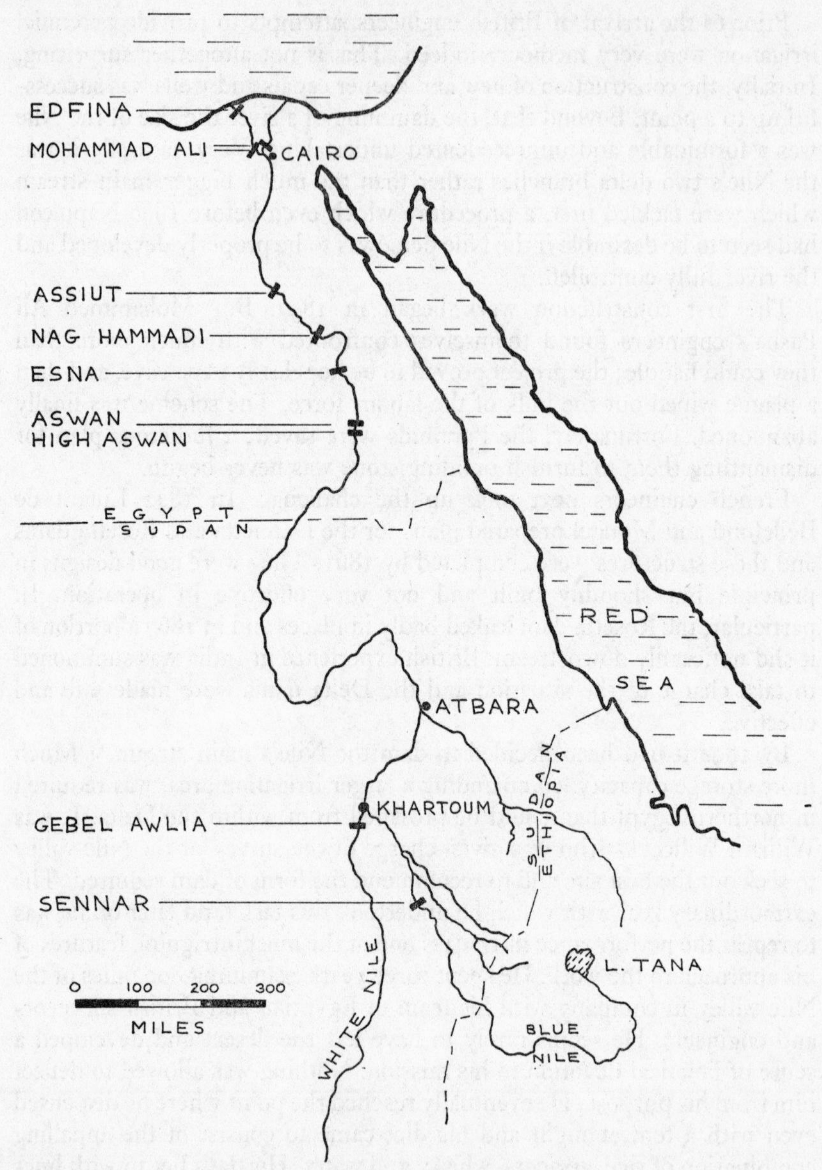

Figure 12 Egypt – the Nile and its dams.

180 deep sluices through which to pass not only water but silt as well, the naturally fertile silt which would promote agriculture if used on the land but choke the reservoir if allowed to accumulate.

Having received the approval of an International Commission, work began on the Aswan dam – and its complementing Assiut dam, 350 miles downstream – in 1898. Four years later it stood finished, a plainly functional structure 6,400 feet long and 65½ feet high holding back when full a lake 200 miles long. It was an epoch-making achievement; the Aswan dam more than any other work marks the beginning of modern irrigation on a grand scale. But for all the benefits which Willcocks' designs and the construction work of Sir Benjamin Baker had brought to Egypt, the country hardly foresaw that what had really been initiated was an unrelenting search for more and more water, and that this would in time throw up some very unexpected problems of far-reaching consequence.

Sir William Willcocks' association with Aswan ended in 1897 and subsequently, for a time, he was engaged on waterworks schemes and drainage projects in Cairo. Through his works and writings Willcocks had become an established international authority on irrigation and in 1908 he directed his attention to Iraq.

As related earlier, Mesopotamia by the thirteenth century was in an advanced state of agricultural decay, the central factor of which was extreme salination brought about by evaporation, transpiration and capillary action over hundreds of years. It is a classic case of over-irrigation leading to disastrous long term results. For six centuries the delta of the 'two rivers' lay waste, a barren and depopulated wilderness by comparison with the fertile eras of Babylon, Assyria, Sassanian Persia and Islam.

At the beginning of the twentieth century Iraq, still called Mesopotamia, was under the control of the Turks who resolved to attempt the revitalization of the region. William Willcocks accepted the responsibilities of a detailed hydrographic and topographical survey which he completed in 1910 after only two years of effort. It was a feat in the same style as his survey of Egypt. An interesting facet of his approach is the extent to which he took account of the concepts of his ancient predecessors, concepts which Willcocks utilized in deciding his schemes, although not always for the right reasons. For the promotion and protection of Iraq's agriculture the traditional issues of water diversion, storage and flood control were fundamental, but compared with the Nile there were special problems: there are two big rivers, not one; the two rivers are prone to unpredictable floods of varying size and at different times of the year; the Tigris–Euphrates delta is very flat, which encourages meandering and also produces serious flooding should the rivers' embankments yield to a freak rise in water level.

To cope comprehensively with these problems Willcocks envisaged control works of two types.[13] In the routine manner he planned barrages to divert water into canal systems based very largely on old established routes. Of the four dams projected only the Hindiya barrage (Figure 13) on the Euphrates was ready in Willcocks' lifetime. It was inaugurated in 1913 just in time to avoid the turmoil of the First World War which was so disastrous for Turkey and her territories. Iraq came under a British mandate in 1920, but not until 1939 was the second of Willcocks' proposals

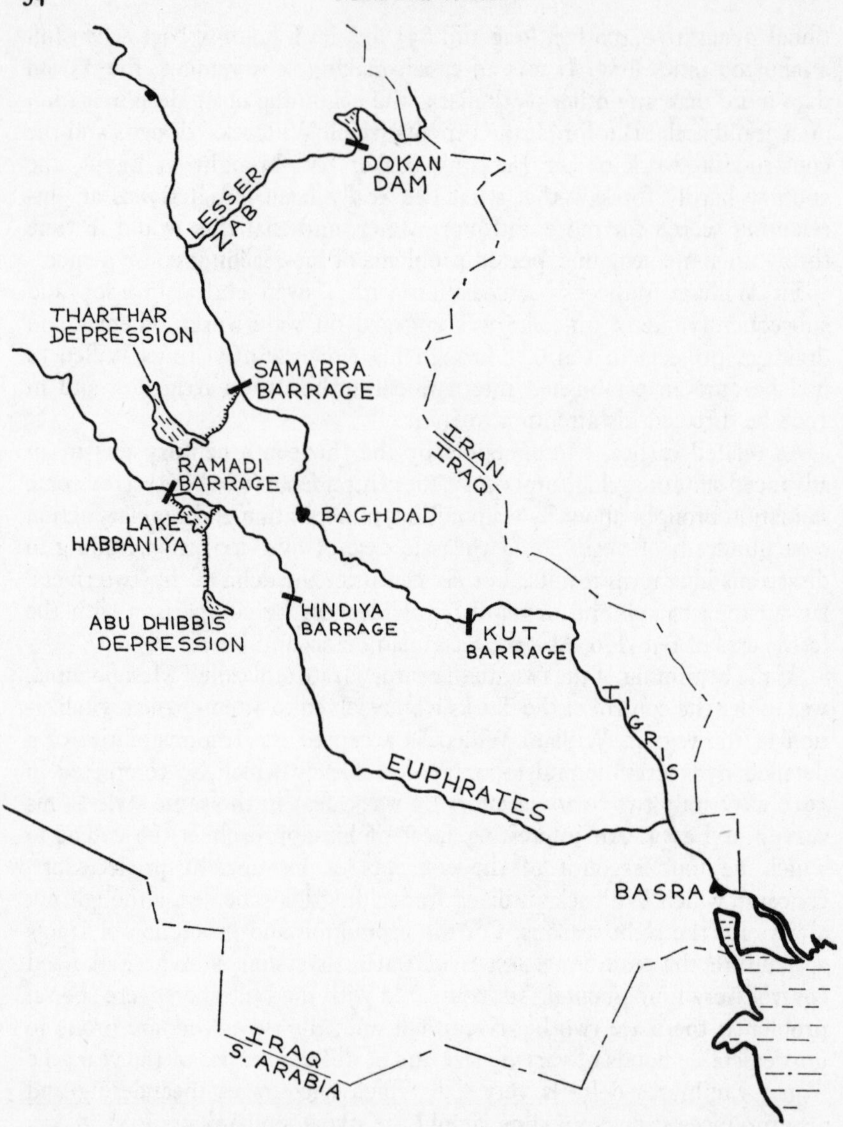

Figure 13 The Tigris and Euphrates and their modern dams and hydraulic works.

brought to fruition. The Kut barrage, a 1,600 foot diversion dam across the Tigris, was completed shortly before the outbreak of the Second World War.

Apart from planning a system of dams to provide perennial irrigation, Willcocks also addressed himself to the potentially more critical question of flood control or what, in the event, turned out to be flood relief. Near both rivers he was able to locate large natural depressions in the desert, the Wadi Tharthar within reach of the Tigris west of Samarra, and the Habbaniya and Abu Dhibbis depressions close to the Euphrates below Ramadi. Willcocks was probably mistaken in crediting ancient engineers with the utilization of these large hollows to absorb extreme floods. But almost certainly when heavy floods *did* break the rivers' banks much of the water would have finished up in the depressions. Anyhow, Willcocks drew up plans to couple the low-level depressions to the Euphrates and Tigris by means of canals and regulators so that floods could be deliberately diverted and absorbed.

In the case of the Habbaniya depression he went one stage further by envisaging a second channel along which stored flood water could feed back to the Euphrates as the requirements of downstream irrigation demanded. The realization of William Willcocks' bold schemes for flood relief and storage was a long time coming. Not until 1956, with the completion of barrages and regulators at Samarra and Ramadi, was it finally possible to claim that Iraq in general and Baghdad in particular were safe from the 'two rivers'.

While the Tigris and Euphrates are the dominant rivers of Iraq, they are by no means the only ones. On its eastern bank the Tigris has important tributaries such as the Greater and Lesser Zabs, the Adheim and the Dyala. To a degree all four had figured in ancient irrigation and Willcocks himself gave some attention to the Dyala. In modern times these eastern tributaries have come into much greater prominence. Their potential for providing not only irrigation but also hydro-electric power is being developed and in the process more flood control is achieved. A notable feature of these developments was the completion in 1958 of the Dokan dam on the Lesser Zab, not only the first large dam and reservoir in Iraq but also a landmark in the evolution of the structural analysis of arch dams.[14] So at last Iraq is being rescued from the oblivion into which it descended for six centuries. Irrigation is being revitalized; drainage works are remedying and controlling the extreme salination which once accumulated but must never recur; and now the benefits of hydro-electric power are being extracted as well.

British irrigation engineering began its development in India, produced some of its finest works there and subsequently migrated to Egypt and Iraq. Absorbed into the repertoire at one stage was the long-standing expertise in these matters of Spanish engineering. The influence of Spain was also felt in the colonial possessions of another European power: France. The conditions, however, were markedly different. Whereas India, Egypt and Iraq were all regions of very big rivers, Algeria was by no means so well endowed, although at least this had the result of making Spain's experience much more directly applicable. And in any case the climates of southern Spain and north Africa are both formally classified under the heading 'semi-arid'.

Algeria was an area where colonial powers had irrigated before. North Africa generally had seen extensive efforts by Roman engineers, the biggest of all their activities in irrigation. Islam did little to resurrect the finest of Roman works – particularly the dam-based irrigation of Tripolitania – but it did prosecute the use of qanats (called *foggaras* in Algeria, *hattaras* in Morocco) on the biggest scale anywhere outside the Persian area of their original introduction.

In the uncertain first thirty years of French influence in Algeria, there was scarcely any interest in irrigation, although in 1844 a Captain de Vauban of the Oran Engineering Corps did design and build a 9-metre high dam across the Wadi Mekkera near St Denis-du-Sig. (The Oran region had been considerably irrigated by the Romans and was to figure prominently in French developments.) Maurice Aymard's investigation of Spanish irrigation technology and its organization marks the beginning of serious French effort. *Irrigations du Midi de l'Espagne* was published in 1864 and work on the Habra dam began a year later. The original intention of constructing an earth dam was abandoned as a result of two local failures of this type of structure, and it was decided to revert to the Spanish irrigation practice of water storage behind a dam of masonry, a straight gravity structure 1,066 feet long and 117 feet high. Complete with an arrangement of typically Spanish outlet and scouring galleries, the Habra dam was operational by 1873.

Other developments followed in quick succession. The Tlelat dam was finished in 1869, Djidionia in 1875, the Gran Cheurfas dam, not far from Vauban's dam, was in use by 1884, and the dam of Hamiz by 1885. A few more were in service by 1900. So in the space of thirty-five years French engineers produced a massive increase in Algerian irrigation backed by over 100 million cubic metres of stored water. Their efforts were not an unqualified success. Rates of siltation were in some cases excessive and not even the provision of Spanish scouring galleries could cope. More serious, however, were structural defects.

As related in Chapter 9 it was French engineers who had, through the theoretical deliberations of de Sazilly and Delocre in connection with the water-supply of Saint-Etienne, pioneered methods of stress analysing dams and introduced canons of structural design. The irrigation dams of Algeria were treated to these two benefits, but the simple concepts thus far formulated were not properly applied, probably because their limitations and the elementary assumptions on which they were based were anything but fully comprehended. The result was two accidents. In 1885, 40 metres of the Gran Cheurfas dam collapsed, but not so much damage was done that the structure could not be repaired. The failure of the Habra dam was more serious. In December 1881 an unusually severe storm exposed a grossly inadequate spillway capacity which allowed the reservoir level to rise 13 feet above its designed maximum. The overloaded profile was thereby subjected to excessive tensile stresses on its upstream face, tension cracks opened up allowing internal water pressure to develop sufficient uplift to rupture 100 metres of the dam right down to the foundations. Six miles downstream the city of Perregaux was badly damaged and flooded (the reservoir, remember, was more than full) and over 200 people were killed.

The lessons of the Habra disaster were not well learned. In 1895 the Bouzey dam in France collapsed for a very similar set of reasons and even more tragic was another failure, in 1927, of the Habra dam itself, by *the same mechanism as had wrecked it in 1881*. Failures such as Habra and Bouzey contributed greatly to the study and perfection of methods of dam design around 1900, a process in which French engineers, not surprisingly, were very much to the fore.

In Algeria dam building for irrigation went on. As in Spain, the aim was maximum development of whatever streams and cultivable areas could be found. Frequently, neither the reservoirs nor the irrigated patches were very large and nothing remotely resembling the scale of operations in Egypt, Iraq or India was feasible. In 1920 plans for eight more dams were prepared and attention was given to the maintenance of existing ones; unfortunately, the latter did not save the Habra dam in 1927. The eight new dams plus three more were finished by 1949, providing 750 million cubic metres of storage to water nearly half-a-million acres. Erosion has been an acute problem and French authorities have put much effort into re-afforestation in an attempt to correct the cumulative effects of too many centuries of tree-felling and goat-herding. Despite all this vigorous activity, Algeria has been no more successful in winning the race between population and food than Egypt or India. In 1957 Pierre Rouveroux observed that 'it would be dangerous and illusory to assume that we can hope for an indefinite rise in output parallel to the indefinite rise in population, which is 200,000 per year'. To which Raymond Furon added the remark: 'An Algerian in 1870 had 5 quintals of cereals per year; today he has only 2.'[15]

5

The United States

SPAIN'S CENTRAL POSITION in the history of irrigation and her role in shaping nineteenth-century developments have been emphasized. Already we have indicated, in Chapter 2, how Spanish influence also travelled west to the Americas and how vigorously the conquerors irrigated not only in Mexico but further north in the arid south-west of the United States as well.

In one sense the Spaniards' position was similar to that of the English and French colonialists in India, North Africa, Egypt and Iraq; irrigation was not new to the regions which were settled and in some cases in fact its origins were very old. Where appropriate, the Spaniards continued to use existing New World installations but for prolonged settlement and long-term development of their new possessions they depended principally on their own experience and expertise.

The dams and irrigation works which the Spanish built up around their missions in the south-western United States were small-scale affairs, highly effective so far as they went, but never intended for more than very localized agriculture. In the nineteenth century we are confronted in the United States by a wholly new facet of irrigation – its relation to the 'frontier'.[1]

Up to the middle of the nineteenth century the United States west of the Mississippi was unknown country, the 'great American desert'. The expeditions of Lewis and Clark, Zebulon Pike and Stephen H. Long had all served to confirm this concept, at least in the minds of people in the East, and the conviction that the area was an uninhabitable and uncultivable wilderness was held to be a fact. Only the experiences of a handful of traders suggested that the Great Plains might be settled and developed, albeit with difficulty and certainly not with the techniques and implements of the humid East.

Around 1850 the West began to be 'opened up'. The annexation of

Texas in 1845, the acquisition of Oregon in 1846, the Mexican cession of 1848 and the Gadsden purchase of 1853 finalized the United States' contiguous continental limits. Now that the 'American desert' belonged to the United States it began to claim more attention. In the first place, it was a route to the developing Pacific coast, itself now a part of a nation which had been reading since the 1830s lengthy reports and descriptions of the territory beyond the Rockies. Ideas for trans-continental railroads quickly sprang up and produced a sequence of voluminous reports. Then there occurred two westward migrations of great significance.

The discovery of gold on Sutter's Creek in January 1848 saw thousands 'rush' to California by sea, by southern routes but mostly by way of the California trail through Nebraska, Wyoming and Nevada. The subsequent quickening of California's development – recognized by the conferring of statehood in 1849 – crystallized plans for railroad construction. The Union Pacific route became a reality in 1869 and four more railroads were opened by 1884. The railroad transformed the West by providing communications and transport. When a decline of California gold mining set in, which came about quite quickly, the railroad allowed and encouraged a shift in activity to Nevada, Arizona, Idaho, Montana and Colorado. Just as important to the West in the long run was the development of cattle raising and the cultivation of wheat, cotton and citrus fruits. Water resource developments, particularly irrigation, were fundamental to the establishment of these industries; and the railroad moved the produce to its markets.

The other important migration of 1848, of fundamental relevance to our theme, was the first large influx of Mormons into Salt Lake in Utah. Their purpose was to establish an independent society free from religious persecution. But to survive in such an arid environment recourse to irrigation was essential. With hard work, shrewd planning and carefully controlled use of whatever meagre water resources could be located, the Mormons showed that the desert could be cultivated. Therein lies the importance of the first large-scale irrigation ever undertaken in North America; it was shown that it could be done. The exploit marks the beginning of many developments.[2]

When the Gold Rush hit California, the most immediate and pressing hydro-technical problem was power, as Chapter 14 of this book describes. But irrigation problems soon forced themselves upon California's fast increasing population, particularly in those numerous mining areas which were remote from food supplies or too arid to produce their own. Elementary irrigation (and power too) was achieved by diverting streams, and a few surface wells were dug, first of all for domestic purposes and watering animals. Artesian water was also located but extended cultivation quickly over-taxed all supplies of underground water, especially after the introduction of pumping. With the fairly rapid decline of gold production in California, the mining emphasis shifted to silver and copper and there was a general migration of miners back over the Rocky Mountains to new areas of activity. California itself began to search for new enterprises as a basis for economic development. In the 1870s two brothers, George and W. B. Chaffey, recently arrived from Canada, began to explore the potential of fruit growing and dramatically indicated California's potential, *given irrigation*. But there were two principal defects in existing Californian irrigation practice, such as it was, which stood in the way of developments: existing water-law and lack of water storage.

Lacking any established system of water-rights, with no precedent to turn to, and in the absence of professional lawyers, Californian miners had resolved questions of water usage by the time-honoured process of rigging up sets of rules acceptable within a given locality. Basically they evolved a system of appropriation which accepted that water was diverted more or less permanently. After use – for washing, fluming, power production and so on – there was no obligation to return the flow to its parent stream. Such a doctrine of appropriation was fundamentally a contrast to the riparian doctrine long traditional in Northern Europe and New England.

Riparian water-law allowed the use of river water only by the owner of land adjacent to the stream and only on his *riparian* lands at that. All riparian owners had this right, whether they wished to exercise it or not, and all water once utilized, in a mill for example, was returnable to the river so that the flow was undiminished overall. In arid and semi-arid lands this doctrine would not work; the Gold Rush proved the fact in California, and the Mormons ran up against the same difficulty in Utah.

The system of appropriation evolved by the Mormons presupposed that whatever water resources were available should be used on any productive land, wherever it was located, for the general benefit of the community. Private water rights and personal advantage were rejected in favour of public ownership and beneficial use. The equitable administration of the system and the punishment of offenders who misused or wasted water was overseen by appointed water-masters. There is, of course, much in these attitudes which parallels irrigation practice in parts of the Old World and Spanish-America, although this cannot at present be taken to show that earlier traditions in any way shaped Mormon thinking. The same must be said of the irrigation technology itself. Natural watercourses were diverted into a main canal which in turn fed secondary canals, or 'laterals', and from them 'corrugates' brought water to the plants. Ditches led water from crops at one elevation to those below; the arrangement is familiar.

The appropriation doctrine, crudely evolved by miners and fully developed by Mormon irrigators, established itself as the basis of arid zone water-law. Its advantages are generally recognized and by and large it continues to be used to this day.

Given a framework of rules to govern water usage, California's other problem was the acquisition of as much water as possible. The extension of irrigation to new and larger areas and to the cultivation of new crops, especially grapes and citrus fruits, required storage. The impoundment of rivers behind dams had not up to this time, *c.* 1880, been much practised in the United States, east or west, for any purpose, and the publication of William Hamilton Hall's *Irrigation Development: History, Customs, Laws and Administrative Systems Relating to Irrigation, Watercourses and Waters in France, Italy and Spain* in Sacramento in 1886 is an indication of the search for ideas. Two years later the first of many editions of Edward Wegmann's *The Design and Construction of Dams* placed masses of imported information at the disposal of American dam engineers; some three dozen dams in Algiers, India, Italy, France, Spain, Belgium and Britain are detailed. By contrast Wegmann's list of American dams is noticeably short.

The first dam built in California to impound irrigation water reflects in every way the prevailing 'frontier' situation: inexperience, fresh ideas about dams, economy of construction, hasty execution and nerve. The 'eighth wonder of the world', James Dix Schuyler called it.[3] Early in the 1880s the Bear Valley Irrigation Company of Redlands decided to make better use of the Santa Ana river in the San Bernardino Valley by building a storage reservoir. The dam site, good in itself, was remote and difficult of access: 6,750 feet above sea level and 70 miles away across desert and mountains. To reduce both the cost of materials and the problem of transportation of materials and equipment to a minimum, F. E. Brown, the engineer in charge, elected to build an arch dam of strikingly slender proportions. At this time thin arch dams were notable for their rarity. Modern research has revealed but a handful of examples which pre-date 1880, located in Spain (Elche, 1642), France (Zola, 1854), Italy (Ponte Alto, 1613) and a few in Persia; of these, few if any could have been known to Brown.[4] So the designer, with economy and speed more in mind than structural accuracy, used the elementary cylinder formula, in conjunction with design stresses of 620 lbs. per square inch, to produce the unprecedentedly daring design shown in Figure 14.

The Bear Valley dam, completed in 1884, attracted the apprehension of numerous engineers throughout its life. But in fact for twenty-six years, and in spite of a dangerously inadequate spillway, the structure performed without trouble; indeed it was a huge success.

In 1902 James Schuyler wrote of the Bear Valley dam:

> The creation of the Bear Valley reservoir has more than doubled the area of land irrigated previous to its construction in the territory covered by its water, and has increased the valuation of property in far greater ratio. The useful function of the storage reservoir was never more fully exemplified than in this case.[5]

And so it was. The Bear Valley dam not only demonstrated the immense value to Californian irrigation of water storage but also it suggested the feasibility of arch dams. Certainly F. E. Brown was satisfied. In 1886 he planned a similarly bold dam on the Sweetwater river near San Diego but local irrigation interests insisted that this much bigger structure be more amply proportioned.

In 1895 the Sweetwater dam successfully survived a massive flood lasting for two days (not unusually its spillway capacity was grossly inadequate). Not only was this pounding a testimony to the quality of the dam's

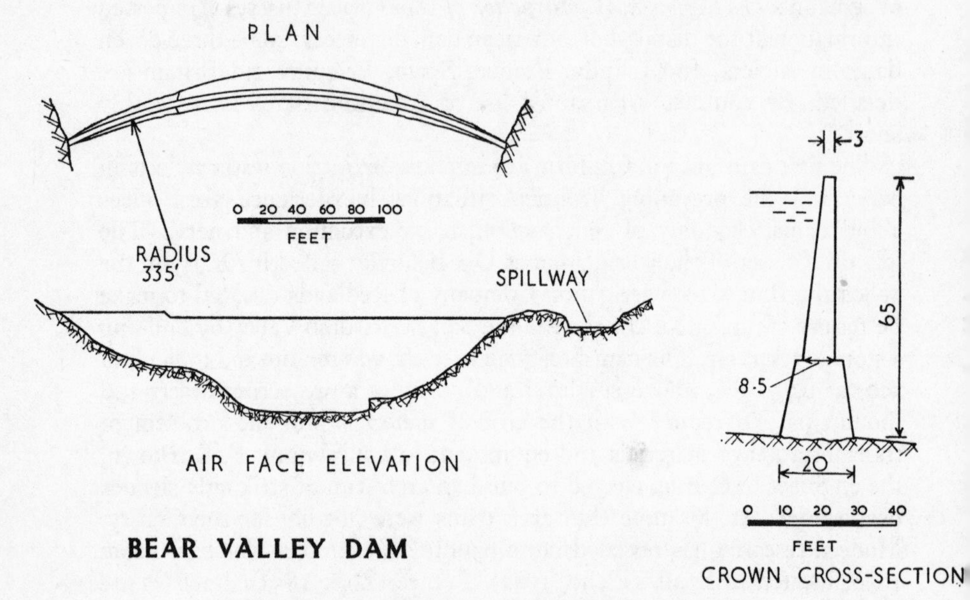

Figure 14 F. E. Brown's daring arch dam of 1884.

construction, but it firmly established that arch dams were a sound proposition. The principle was increasingly applied to dam construction in California, e.g. the Hemet dam, 1895; the La Grange dam, 1894; and the San Mateo dam, 1888. At the beginning of this century a crop of thin arch dams sprang up in Australia.[6] Here too the emphasis was on economy of material, and in both the United States and Australia this same quest for cheap construction was increasingly met with concrete, the San Mateo dam being the first to be built entirely of this material.

Storage dams for irrigation, first exploited in California, quickly became part of the irrigation scene in other states, notably Arizona, Montana and Colorado. In the first twenty years the development of irrigation in the West was in the hands of private irrigation companies. So far as they went, these efforts were successful and emphatically indicative of what could be done. The necessary technology was rapidly developing and the doctrine of appropriation was a workable system of water distribution. Nevertheless, the development of irrigation for the benefit of the whole country required organization and planning on a grander scale than private investment could ever provide. An effort to broaden the basis of agricultural developments had been made in the 1894 Carey Land Act but the effect was minimal. Individual states were unable to exercise their rights to develop water resources due to lack of money and conflicts between regional needs. Ultimately, the solution was seen to lie with the Federal government itself.

By conducting both geological and hydrological surveys, the Federal government had already been involved in the affairs of the West with valuable results. More direct and concrete participation was initiated with the passing of the Reclamation Act of 1902. Private and corporate enterprise was by no means superseded as a result, but for all that the real irrigation power in the land was now the state. It is perhaps worth noting that this situation on a comparable scale had hardly existed anywhere since antiquity and with the formation of the Bureau of Reclamation the state's role was explicitly recognized.

For a few years following the Act of 1902 the principal effort was directed to surveys of the location and scale of irrigation needs and to the identification of favourable sites for dams, canals, pumping stations and control works. The Bureau's first practical exercise in irrigation was the Truckee-Carson project, begun in 1903 in western Nevada. The first dam built by the Bureau—a 20-foot high diversion dam on the Carson River – was a modest structure in itself but a prelude to civil engineering in the grand manner. Arthur Davis' *Irrigation Works Constructed by the United States Government* of 1917, and later editions of Wegmann's *Design of Dams*[7], chronicle the amazing rate at which dams sprang up (and with them extensive structural, mechanical and electrical ancillaries), and among them is a sequence of structures and techniques of the first importance. Following the precedent set in California, the Bureau's first two big dams were arch structures, the Pathfinder (1907) and the Shoshone (1910) dams in Wyoming. Their trapezoidal profiles were geometrically similar, and at 328 feet in height the Shoshone dam was substantially the highest in the world. The great significance of both lay in the method of their design. F. E. Brown had used the fundamentally inadequate cylinder formula in his analyses, as had Australian engineers for their arch dams. In order to represent better the structural action of an arch dam, Bureau engineers introduced the concept of a co-existing system of horizontal arches and vertical cantilevers. The design process was based on the notion that the water load on the dam could be divided between the two systems of structural elements in such a way that both underwent the same rotational and translational movements at their points of intersection. For Pathfinder and Shoshone the analysis was restricted, for ease of computation, to a few arches and one cantilever, the one at the crown.

The so-called 'trial-load method' put arch dam design on a much sounder footing, particularly if a series rather than one cantilever was used, and especially if complete compatibility of movements at all points was achieved. The technique has remained fundamental to the American school of arch dam design to this day and has been the method used to design many large and famous dams: the Arrowrock dam (400 feet high) on the Boise river in Idaho; the Gibson dam (200 feet high) in Montana; the mighty Hoover dam (727 feet high) in Nevada; and the Hungry Horse dam (564 feet high) in Montana.

While the Bureau of Reclamation's initial purpose was agricultural development, its functions soon extended to fields other than irrigation and its influence was felt well outside the United States. 'Not even Roosevelt,' writes Carl Condit, 'could have imagined the enormous scope of the Bureau's operations a half-century after its establishment.'[8] Quite apart from the attention given to arch and arched dams, which was of such international interest that it prompted research elsewhere and led eventually to alternative design philosophies, other types of dams were studied; the buttress dam was exploited under favourable conditions and some notably large examples were built, e.g. the Coolidge dam, the Stony Gorge dam and the Bartlett dam. Of the more conventional gravity type, the Grand Coulee and Shasta dams are still among the world's largest. In a response to three factors – cost, the need to exploit certain types of site and the potential of modern machinery – Bureau engineers have contributed much to the increasing use in modern times of earth and rock dams.

It would be idle and unfair to leave the impression that the U.S. Bureau of Reclamation has been the only important influence in modern dam building and irrigation. Of course many other agencies in the Americas, in various European countries, and more recently in Russia and China, have made notable contributions. Nevertheless, the Bureau's role has been central, particularly between 1900 and 1950, and has typified a variety of modern trends such as the multi-purpose hydro-scheme.

Water storage on a grand scale, the scale on which the Bureau of Reclamation chose and indeed was expected to work, represents massive investment of money and equipment. At first it is tempting and ultimately it is vital to think in terms of more than one application of the water and more than one use for the reservoir. Water stored for irrigation can often be conveniently used to generate electrical power with no disadvantage to the irrigation sequence (and in fact some power is usually essential to the irrigation itself to power pumps). In 1911 the Roosevelt dam, primarily for irrigation, became a supplier of electrical power to Phoenix, Arizona. The policy was developed and refined until multi-purpose objectives were integral features from the outset of planning. Power and irrigation needs are both met by the Hoover dam and the Columbia basin project[9] and in other schemes public water-supply is provided as well.

Multi-purpose schemes utilize dams not only to store water for use but as a means of river control as well. The United States has more than one serious flood problem of long-standing concern, notably that of the Mississippi valley. Early in the eighteenth century a New Orleans engineer called De La Tour initiated the first flood control works on the river and at intervals for the past two centuries more and more has been added in the way of levees, flood relief channels, flood escapes into natural reservoirs and flood control dams. Such dams present possibilities for power generation or irrigation; they may also present obstacles to navigation and movements of fish. So the effective use and control of all aspects of a river basin's potential by means of multi-purpose dams is both conceptually complex and operationally exacting.

In 1933, the year when the project to develop the Columbia River and its tributaries was begun, the most elaborate attempt so far at regional revitalization and development, based on an integrated system of multipurpose dams, was undertaken. The aim of the Tennessee Valley Authority was to control and harness the Tennessee river and its tributaries in the interests of navigation, flood-control, hydro-power, soil conservation, irrigation, fishing and public health. Quite apart from the technical problems involved and the need to construct more than thirty dams, together with a wide range of ancillaries, the task of the TVA entailed problems of finance, organization and co-operation on a grand scale. Not always is it an easy matter to meet the conflicting requirements of power-generation, flood control, navigation and malarial extermination in such a complicated system spread over a large territory.

The problems of multi-purpose irrigation and river control systems encountered by the USBR and the TVA prior to the Second World War have since proved similarly onerous in other projects. Equally, economic pressures and the needs of underdeveloped countries, where improvements are expected to accrue rapidly, have led to an increasing dependence on multi-purpose hydro-schemes. In Africa, India, South-East Asia and other places, large irrigation works are rarely planned nowadays in isolation; some additional benefits, power being the most usual, are always realized. However, the idea is not without its limitations. In power-plus-irrigation projects, the hydro-electric phase must precede crop watering; dams upstream, crops downstream. But the necessary geographical layout does not always exist. Mexico, Spain and East Africa are good examples of regions where the bulk of the agricultural potential lies inland on high plateaux while the most favourable hydro-electric sites are coastal.

For several thousand years, land hydraulics has figured large in the role and development of civil engineering; and it still does. In the interests of land usage, health and agriculture, land drainage and reclamation projects are frequently undertaken, and even more important is modern irrigation. In both the developed and underdeveloped areas of our planet, more and more land is being watered in the hope that food production will win the race against population increase. It is understandable that the traditional concepts of dam building for water storage, of impounded water feeding canals, of canal systems distributing water over the land and of drains running excess moisture away should continue to be adhered to. After all the technology is well understood, and improving, while the results can be guaranteed.

Nevertheless, opinions have been expressed that in the future irrigation may not follow the course which the study of its history suggests as logical. 'Is Large-scale Irrigation Reaching an End?' is the question posed in Chapter VII of Georg Borgstrom's *Too Many*.[10] The nature of Borgstrom's pungent and critical answer will be examined in our concluding chapter, *Man and Water*.

Part II

Water for Drinking

6

Wells and Springs, Qanats and Canals

A SUPPLY OF POTABLE water, certainly in developed countries, is probably the vital service which above all others we take for granted. The possibility that nothing will happen when a tap is turned or a lever pushed is one that quite simply we never consider. In times of war and other emergencies – severe disturbances in the weather, for instance – vigorous efforts are made to maintain a water-supply because the consequences of any failure to do so would be catastrophic.

Aside from oxygen, water is man's most important requirement and without a daily supply of at least one and a half pints his life expectancy is only a matter of days. Lack of food is only a threat after several weeks, and in a reasonable climate such basics as shelter and clothing can be dispensed with altogether. The purity of the water we drink is important too. Bacteriological impurities can be as deadly to health, and ultimately life itself, as no water at all. It is true to say, nevertheless, that with one notable exception public water-supply is very much a modern achievement hardly more than a century old. Before that virtually all attempts to establish public water-supply systems were more or less deficient owing to various technical, social, medical and organizational shortcomings.

Although the essential components of water-supply technology, including even the concept of bringing water to settled communities, have very early origins, man's efforts to procure a personal supply are very much older. However, this very early phase, really a part of pre-history, can be dispensed with briefly since it is pre-technological.[1]

Good drinking water occurs naturally in rivers, lakes, springs, oases and underground. Primitive man set himself no higher objective than the location of such sources and mentally noted their position. Even today, some very primitive societies still function at this most elementary level and their ability to locate sources of water is uncanny. Finding surface water in certain types of terrain and under some climatic conditions is usually relatively straightforward, but not always. The incidence of springs and water-holes in arid and desert lands is rare and it remains uncertain how early man plotted his course from one point of supply to another. Underground supplies are even more of a mystery. It has been suggested that well-digging was the result of efforts to enlarge the supply from natural springs, but that argument is ultimately unconvincing because most wells have been dug at places where no visible outflow was possible. Conceivably, the driving of shafts and drifts in the pursuit of metallic ores gave some guidance but in the final analysis one is bound to suppose that well-sinking

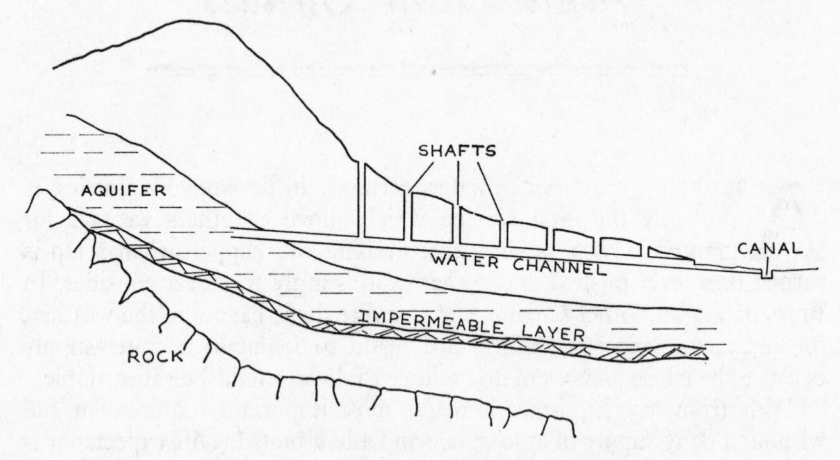

Figure 15 The layout of a typical qanat.

was wholly speculative. People must have dug for water where they needed it, with no guarantee of a successful outcome, and many a disappointment.

Eventually, the occurrence of good supplies of drinking water, and the establishment of the means to reach it, was an important factor in the location of the earliest settled communities. The sinking of wells represents a facet of the emergence of a technology of water-supply and as the demands of settled groups of people grew, so did the dependence on specific techniques. In ancient times there were four distinct areas of technical development: location of supplies, mechanical water-raising, water conduction and storage.[2]

Taking for granted the utilization of surface water, the most interesting aspect of ancient water-resource technology is an extension of the concept of wells to a novel technique for locating and utilizing underground water: the qanat. Essentially the qanat is a horizontal well (see Figure 15). In hilly country a slightly sloping tunnel is driven from some suitably elevated point into the hills which are known, or hoped, to contain an aquifer.[3]

Qanat construction represents, along with certain types of mining operation, the first serious activity in the difficult and dangerous enterprise of tunnelling. Indeed, in origin, mining and qanat building may well be very closely linked technologies. It is nowadays accepted that the spread of the qanat system was initiated by the Assyrian king, Sargon II (722–705 B.C.), following his campaigns in north-west Persia, or what is now Armenia. The same region is one of the oldest metal-mining centres of antiquity, so the interrelation is plausible. In addition, the tunnelling procedures involved were similar.

Paradoxically, the fundamentally horizontal qanat was engineered by means of vertical shafts. At intervals of about 50 metres on a line running towards the aquifer, vertical 'wells' were dug to an appropriate depth. By joining the bases of these shafts the near-horizontal adit was built up. The system has the advantages that many sections of the work can proceed concurrently, outlets for the tunnelling spoil are automatically provided and subsequently, when the qanat is in flow, the vertical shafts provide both access and ventilation. Where the qanat is very deep, that is to say in the furthest portion which strikes into the aquifer, then the vertical inlets were often dispensed with.

At the time of its inception, qanat building was a relatively advanced technology. It was, moreover, one which made demands on planning, organization and calculation as well as physical effort. But given the means and the will to carry out the construction, qanats represented a good investment with substantial long-term advantages. A good aquifer would yield a reliable and steady flow from year to year and the water was generally of good quality. Conduction of the supply along a tunnel aided cleanliness and offset to a large extent the problems of evaporation associated with lakes, open storage and open channels.

Overall the advantages of qanats are attested by the extent to which their use spread, and their sustained utilization right down to modern times. Originating in Armenia, their use became widespread throughout Persia and then spread into Syria, Arabia and Egypt. About 300 B.C. Megasthenes reported many examples in northern India and commented on the procedures evolved for their administration and supervision. Roman influence extended the use of qanats along the North African coast, and later on Islamic expertise continued the diffusion as far west as Morocco and also northwards into Sicily (at Palermo), and to numerous places in southern Spain. A new phase of transmission occurred when Spaniards introduced qanats to Mexico and other parts of America.

Qanats, some of considerable age, are still widely used and occasionally new ones are constructed, especially in Persia, the area of their origin. As recently as 1933, the city of Teheran drew its entire water-supply from a system of qanats.

The application of simple devices such as ropes and buckets, shell dippers and the shaduf (see Figure 1, page 8), to the raising of water from wells represents the earliest of all efforts to 'mechanize' water-supply. None of these devices, however, are suitable for anything but very small lifts. For deep wells, therefore, the windlass was resorted to and despite a very limited performance its use became universal; it has remained so until modern times.

The earliest recourse to animal-power for water-drawing utilized oxen

Figure 16 From a relief in the British Museum comes this earliest depiction of the use of the pulley. A besieged town is about to be deprived of its water supply.

or camels marching down ramps and this innovation depended on the introduction, in some form, of the pulley. The pulley's origins are obscure. Its first authenticated application occurs in an Assyrian relief of the eighth century B.C., but the source of motive power, man or animal, is not specified (see Figure 16).

In order to supply relatively large quantities of drinking water from deep wells, Roman engineers occasionally made use of the saquiyah, the only water-raising machine in their repertoire whose configuration was suitable for large lifts in confined spaces. The earliest archaeological evidence of this innovation is contemporary with the description of Vitruvius; some examples at Pompeii were man-powered and in a single Egyptian case the saquiyah was driven by animals.

Because the quantities involved were significantly smaller, water-supply in antiquity did not demand mechanization to anything like the same extent as irrigation, and with few exceptions only the most elementary machines were commonly used even in deep wells.[4] The survival of examples of ancient wells fitted with systems of steps, ladders or a series of platforms is evidence of no mechanization at all but rather a continuing reliance on laborious manual methods employing individuals or human chains.

When it came to conducting water-supplies over long distances by means of open channels, the ideas of ancient engineers were doubtless shaped by the experience already gained on a large scale in irrigation. The need to channel water-supplies from distant sources was an outcome, in many cases, of the growth of urban communities. As noted earlier, water-supplies were often instrumental in fixing the sites of primitive settlement. But as the demands of such communities grew, so water had to be brought in from further afield. Alternatively, it was other considerations – strategic, commercial, administrative – which fixed the location of towns, and then the question of long distance conduction had to be faced from the outset.

Long water-supply channels are implicit in the use of qanats. Quite apart from the underground section, sometimes long in itself, the stream also had to be channelled from the aquifer-bearing hills to the delivery point in town or city. The required flow was always gravitational and this required the establishment of a suitable gradient fixed by certain upper and lower limits. The range seems to have been interpreted pretty liberally. Very steep gradients, certainly over any distance, were avoided to prevent erosion of the channel and other damage; very shallow gradients were useless since they reduced the rate of flow and encouraged stagnancy. In between lay a practical gradient, dictated by the lie of the land, the availability of contours, and the nature of the materials used to build and line the water-channel.

The materials of construction varied from place to place and the choice was governed by conditions and availability. Sometimes the channel could be cut straight into the ground without an artificial impervious lining. Otherwise one was provided in the form of a brick, clay or mortar facing. In a few places wood was used, even to the extent of occasionally building a wholly artificial wooden trough which was not sunk below ground level at all. The use of masonry seems not to have been widespread before Roman times.

Whatever the source – qanat, lake, spring or river – an open channel had the advantage over a closed conduit that greater flows could be handled. However, this advantage could only be realized by maintaining the gradient within the prescribed limits. Ancient engineers often went to considerable lengths, literally, to follow suitably gentle slopes along natural contours. But how did they deal with the obstacles presented by high ground, river crossings, depressions and ravines?

If hills could not be economically skirted, tunnels were cut. Once more the close interrelation between mining and water-supply is evident. However, setting the line of such a tunnel and maintaining the gradient was sometimes an exceedingly difficult problem which taxed ancient surveyors and their methods to the utmost and more often than not defeated them entirely. If a tunnel's maximum depth below the surface of high ground was sufficiently small, the qanat technique was viable. From a straight line laid out across the top of the hill vertical shafts were dug to a predetermined depth and their bases connected. The Mount Pentelicus aqueduct built by Greek engineers for the water-supply of Athens used this method; and later on Roman tunnels were built on the same basis.

More difficult was the driving of aqueduct tunnels without the aid of vertical shafts. Recourse then had to be made to geometry, surveying and a good deal of dead reckoning. In order to hasten completion of the work, tunnels appear to have been driven from both ends at once; usually the central rendezvous was only achieved by judicious adjustment to both line and gradient at the last moment. In the case of the famous Greek aqueduct tunnel at Samos, so praised by Herodotus, the engineer in charge, Eupalinus of Megara, had to resort to a 16-foot cross cut in order to make the connection.[5] The Siloam tunnel, built about 700 B.C. in Jerusalem, was angled so many times that the 366 yard distance produced 583 yards of tunnel. The same thing happened at Megiddo in a tunnel whose total length is a mere 65 metres. Much later (A.D. 152), in the case of a Roman aqueduct tunnel at Saldae in Algeria, the failure of the two sections to meet is described by the engineer, Nonius Datus, in his own despondent words.

> But I succeeded in reaching Saldae, where I met the army commander. I found everybody sad and depressed. They had given up all hopes that the opposite sections of the tunnel would meet, because each section had already been excavated beyond the middle of the mountain. As always happens in these cases, the fault was attributed to me, the engineer, as though I had not taken all precautions to ensure the success of the work. What more could I have done? For I began by surveying and taking the levels of the mountain, I drew plans and sections of the whole work, and these plans I handed over to Petronius Celer, the Governor of Mauretania; and to take extra precaution I summoned the contractor and his workmen and began the excavation in their presence with the help of two experienced gangs, a detachment of marine infantry and a detachment of Alpine troops. What more could I have done?

> After four years' absence from Lambaesis, expecting every day to hear good news of the water at Saldae, I arrive; the contractor and his workmen had made one mistake after another. In both halves of the tunnel they had diverged from a straight line, each veering to the right, and had I arrived a little later, Saldae would have possessed two tunnels rather than one.

What more could he have done? Well Nonius Datus could have visited the job more frequently than once in four years. As it was, he took the standard emergency action of building a cross-gallery.

When ancient engineers were forced to take water-channels across depressions at least they could see where they were going. The technique most frequently used to maintain a level grade was the construction of an embankment. Generally they were made of earth or brick or both; occasionally they were of masonry. In order to cross streams or wadis, such embankments were pierced with openings and thereby took the form of a low bridge. Aqueduct bridges of any size were rare, however, and nothing comparable to those built by the Romans have been found.

Quite the most interesting feature of all pre-Roman aqueducts was the occasional use of pipelines to carry the supply below the hydraulic gradient and therefore under pressure (see Figure 17). The term 'inverted siphon' is used to describe the pressurized pipelines which crossed valleys and depressions.[6] Although the use of inverted siphons, especially by Greek engineers, is well established, it is by no means clear how it was done and in particular what materials were used to contain pressures which sometimes were as high as 200 lb per square inch. For small-scale and localized water distribution closed pipes were extensively used in antiquity and archaeology has unearthed many examples. Generally they are of very small size – small in diameter and in very short sections, the quality of their manufacture is varied, and the materials of their construction include earthenware, clay, wood, terra-cotta, stone and lead; examples even of copper and bronze pipe have been found. Opinion is divided as to which material or materials the Greeks may have used for inverted siphons. If the later Roman examples be a guide, they were made of lead, but wooden or terra-cotta pipes are possible. Whatever the solution, pipes of small internal diameter and great wall thickness must have been used, so that only a multitude of parallel pipes could have enabled any siphon to carry the sort of substantial flow which in an open channel was a matter of course.

Impoundment of water at the source of a supply system, a basic proposition in modern installations, was not a feature of ancient systems barring a few rare exceptions. Nevertheless, water-storage was widely practised and for three basic reasons: to make maximum use of the available flow both day and night, to compensate for seasonal variations, and to provide a town supply in the advent of a siege. So far as capacity

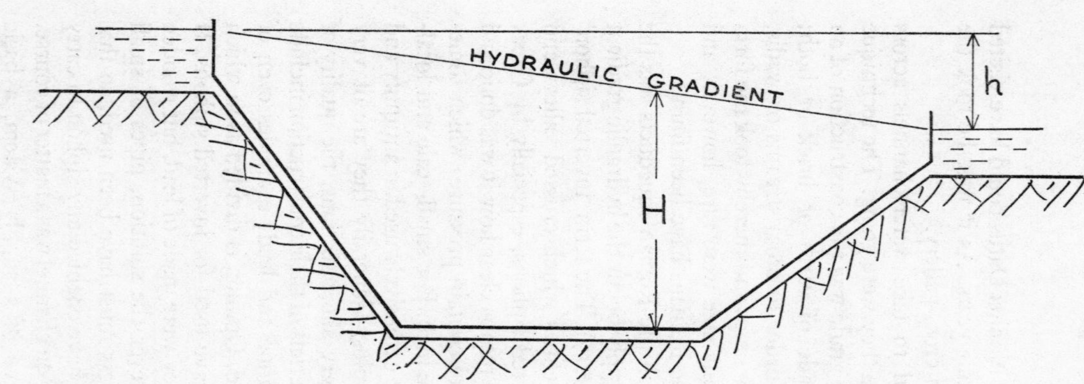

Figure 17 The inverted siphon. The head lost in overcoming friction is given by h. The pressure in the pipe is given by H.

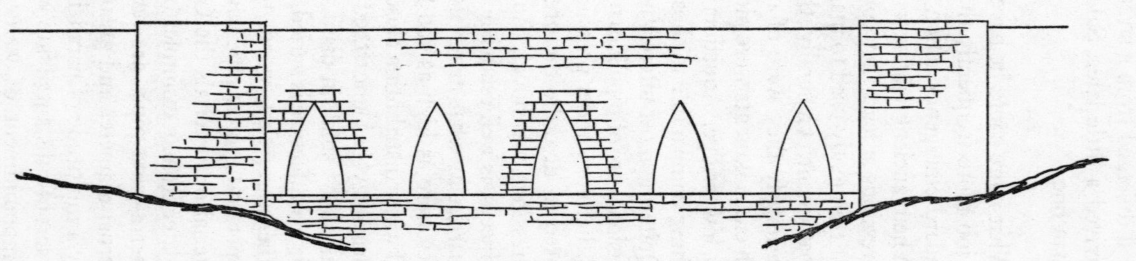

Figure 18 Sennacherib's Jerwan aqueduct carried on five corbelled arches.

was concerned, the last was the most demanding proposition of all. Almost without exception, all forms of cistern and storage tank, from the smallest to the largest, were covered. This helped to keep the water clean and greatly reduced evaporation. So far as mechanical impurities were concerned, any degree of storage aided purity by allowing settlement of silt and suspended particles.

Ultimately, the most interesting facet of the history of water-supply in any age is the establishment of large systems, embracing some or all of the techniques outlined above, and sustaining large populations in towns and cities. Set out below are a few of the most elaborate examples dating from before Roman times with details of whatever features are particularly notable.

Of all ancient cities, Jerusalem can boast extensive water-supply installations developing over the longest period. The original works dating from about 1000 B.C. are traditionally ascribed to King Solomon; and possibly some of the water made available then was for irrigation as well as drinking. Jerusalem's sources of water were springs south-west of the city and, unusually, the available water was collected in three large, artificial reservoirs before being channelled to Jerusalem. The reservoirs vary in length from 400–600 feet and in depth from 30–60 feet. The combined capacity, around 12 million cubic feet, is significant only if the works are as old as Solomon.

The aqueducts to Jerusalem are a complicated collection, of imprecise date and uncertain origin. Typically, Roman reconstruction and renovations make it difficult to discern the details of the earlier components of the system. One particularly interesting possibility, however, is that there was an inverted siphon on one of the lines; 15-inch stone blocks pierced with a 2-inch diameter hole have been found.

Within Jerusalem, water-storage by means of tanks and cisterns was commonplace. Initially, such reservoirs were contrived from natural formations but ultimately wholly artificial ones were built, their roofs carried on a forest of pillars. What began in Jerusalem on a small scale was later on applied on a grand scale in Byzantium.

A distinctly Palestinian water-supply technique was the *sinnor*, basically the traditional well but with elaborations.[7] External water-supplies were more or less useless to any city that was under siege unless it proved possible to tunnel outwards at water-level to an already existing well. This was the purpose of a *sinnor*. The ones at Gezer, Megiddo and Jerusalem are all very ancient, possibly pre-dating 1000 B.C., although in their final form, with carefully graded tunnels to make the supply flow by gravity into the city, they are somewhat later. The *sinnor* of Jerusalem contains an inscription in cursive Hebraic commemorating the closing of the tunnel. Its dating, *c.* 700 B.C., may very well suggest that the Siloam tunnel was cut in an effort to secure Jerusalem a water-supply in the event of besiegement by the armies of Assyria. Coincidentally the Assyrian king, Sennacherib, was at this very time engaged in extensive waterworks of his own.

Nineveh, Sennacherib's capital city, is an interesting early example of a classic problem in water-supply. How does one maintain a satisfactory level of supply as a city and its population grow? Presumably the Tigris was a source of sufficient volume but useful only if a great deal of effort was put into water-raising. In 703 B.C. Sennacherib began to draw water from the River Khosr; a diversion dam across the river fed water into a 10-mile-long canal which was 'dug with iron pick-axes'. By 694 B.C. demand was exceeding supply. From the mountains north-east of Nineveh, eighteen small rivers and streams were canalized into the Khosr and two new dams were provided so that the much augmented flow of the river could be diverted to the city. For four years these measures were sufficient. In 690 B.C. Sennacherib decided that Nineveh and its gardens needed yet more water and so he embarked on the most ambitious project of all. On the Atrush river at a place called Bavian, more than 30 miles from Nineveh, a deep gorge was dammed with masonry. From these headworks Sennacherib's engineers dug a long, winding canal which after a 36-mile-long journey emptied its flow into the Khosr river. Even today substantial traces of Sennacherib's work can be located. None is more impressive than the Jerwan aqueduct bridge, a structure 300 metres long and 12 metres wide which carried the Atrush canal across a valley.[8] Of great interest is the fact that the bridge's five arches were corbelled, not voussoired (see Figure 18).

The water-supply of Nineveh is significant as the earliest example of a scheme of its specific type. The diversion of rivers by dams, contour aqueducts, aqueduct bridges, and augmenting the flow of one river by draining another into it, are all techniques which are of fundamental importance. They were combined by Sennacherib's engineers into what was, for the time, a notably integrated and elaborate system.

Greek engineering provided water-supply systems for a number of cities. An early example is Mycenae whose spring-fed conduits brought in water which was stored in long, narrow, stone cisterns within the city. Water-supplies in Athens depended at first on a collection of wells and springs within the city. By about the sixth century B.C. the search for new sources went further afield. The channelled supply from Mount Pentilicus, passing at one point through its famous tunnel, was one of several aqueducts to serve Athens.

Of all Greek water-supply systems probably the best known – for so much has been written about it – was the one at Pergamon.[9] It was built, along with the city's other famous institution, the Pergamon library, during the reign of Eumenes II, around 200–190 B.C. Beginning far to the north-east of Pergamon in the Madaras-dag, much of the aqueduct line was built on or close to the hydraulic gradient. But Pergamon itself stood on very high ground, so that to reach the city the aqueduct in its final 2,000 metres had to cross two deep valleys separated by a ridge before climbing some 120 metres to the final delivery point. Both crossings were accomplished with inverted siphons, which so far as can be ascertained from published dimensions of the topography, must have produced internal pipe pressures equivalent to at least fifteen atmospheres.

How successful were the Pergamon siphons is hard to tell. By the end of the second century the city was in Roman hands and Roman engineers for some reason saw fit to replace the Greek aqueduct with a conventional unpressurized conduit of their own. As a result, the higher part of Pergamon, the Citadel, ceased to be supplied with water directly. Other Greek cities which installed water-supply systems also made use of inverted siphons. Evidence for them has been found at Patara, Mylasa, Methymna, Smyrna, Olynthus, Catania, Selinus and Syracuse. A sequence of pierced masonry slabs carried the pressure-pipe, which may have been made from earthenware, or wood or a metal. Conceivably, it was the inability of such pipes to withstand the applied pressure that prompted Roman replacement. Alternatively, the paltry flow of a single small-bore pipe of great length may well have offended Roman ambitions and led to the use of conventional open channels.[10]

F. W. Robins wrote of Greek water-supply systems: 'The activities of the ancient Greek communities, however, though not so methodical and advanced as those of the Roman Empire in the matter, seem nevertheless to mark, so far as present knowledge goes, the beginnings of the organization of communal water supplies.'[11] In fact, as we have seen, other communities before the Greeks had put thought and effort into communal water-supply. Probably Robins was really drawing attention to the fact that Hellenic and Hellenistic civilization procured urban water-supply as a matter of course; it was regarded as a normal component of urban life, not a special feature which, by and large, seems to have been the case previously.

Another observation can be made also. In the past most writers on the history of water-supply, perhaps in their haste to get to grips with the numerous and imposing works of the Romans, have failed to notice an important characteristic of pre-Roman water-supply technology. By Greek times all the basic components of what was needed had been identified. Leaving aside for the moment the idea of pumping water from rivers – an approach which had a great vogue in the seventeenth and eighteenth centuries – there was little in the hydraulics of water-supply technology right through from Roman times to the late nineteenth century which had not been tried, at least in principle if not in scale, by engineers in Babylon, Assyria, Palestine, Asia Minor, Greece or somewhere.

7

Roman Water-Supply

WITHIN THE HISTORY of technology generally, a fundamentally interesting issue, and sometimes an important one, is the question of technical transmission and the diffusion of methods and ideas. In the case of Roman technology, the term 'acquisition' is often more appropriate because as the Roman Empire spread out it came to embrace many societies with technical ability, and to occupy many areas in which specific technologies were well established and thoroughly understood. But given this climate in which the acquisition of technical skills could occur, clear proof that it actually happened and evidence of precisely how it happened is not so easy to specify. However, technical fields in which there very likely was a tradition continuing from captured or contacted peoples to Rome itself can be briefly listed.

Roman architecture shows strong Greek influence; Greek architects were often employed by the Romans and Vitruvius discusses at length Greek ideas of form and proportion. In shipbuilding both Greek and Carthaginian forms of oared ships were adopted by Roman navies. Greek military machines and war engines were easily assimilated by Roman armies. Although the point continues to be contested, on balance it seems likely that the semi-circular arch, such a distinctive feature of Roman building, was of Etruscan origin. Much Roman road-building utilized existing Etruscan routes and tracks; however, Carthaginian influence in techniques of road-paving is now discounted. Irrigation techniques adopted by Roman cultivators were long established and fully understood throughout the empire. The tradition in hydraulic machinery, both for lifting water and for power, is particularly obscure but some elements were certainly pre-Roman or extra-Roman in origin.

The most important feature of Roman construction was the use of hydraulic concrete. That natural cements could be made hydraulic by adding volcanic ash – actually Santorin earth – was known to Greek builders. Surveying problems were tackled with instruments and procedures that were basically Greek and frequently the Romans employed Greek personnel as well. In fact, a number of modern writers have stressed how generally widespread was the employment of Greek engineers and how late it persisted. So far as written evidence is concerned the central document is Vitruvius. The *Ten Books on Architecture* draws so heavily on Greek authority that it is as much a treatise on Greek engineering as Roman. As a picture of wholly Roman attitudes to civil engineering, the later work of Frontinus on Rome's water-supply is much more revealing. So impressed was Frontinus with the scope and importance of Roman public-works, that he is moved at one point to state: 'With such an array of indispensable structures carrying so many waters, compare, if you will, the idle Pyramids or the useless, though famous, works of the Greeks!'[1] It is an odd remark. The Greeks, as outlined above, had in fact succeeded commendably in the very field which Frontinus regarded as being so beneficial to urban living and such a mark of civilized development. The earliest Roman public water-supply schemes may well have been modelled on existing Greek installations in southern Italy, Greece and Asia Minor.

For the historian, the water-supply works of Rome itself have always been of major interest. The sheer scale of the system in its final form, the extensive and imposing remains,[2] and the survival of Frontinus' personal account of the installations' administration, are the basic reasons for this fascination and pre-occupation.[3]

Frontinus begins *De Aquis* by recording that 'from the foundation of the City, the Romans were satisfied with the use of such waters as they drew from the Tiber, from wells or from springs'. Ultimately, these supplies proved to be either insufficient or in some other way inadequate or both. Precisely what prompted the first long-distance aqueduct is uncertain, but most likely it was a combination of the problems of purity and quantity. During the fourth century B.C., Rome grew rapidly in size. It is plausible that the greater concentration of people within the city's walls not only over-taxed the supplies from 'the Tiber, from wells or from springs' but that it also brought pollution of both river- and ground-water resources to an unpleasant, possibly a dangerous, level. At the time, the systems of drains and sewers which Rome was later to develop extensively were already sufficiently used to have a deleterious effect on the Tiber, especially in the summer when the river's flow was not notably large.

Constructionally, Rome's first aqueduct, the Aqua Appia built in 312 B.C., was not unlike a sewer, being of small cross-section with a low arched roof. Except for a short distance inside the city, the full ten miles was laid underground. It should be emphasized here and now that this was generally the case with all Roman aqueducts, at Rome or anywhere else. So far as was practicable, the lines of aqueducts followed a steady grade at or below ground level. The use of tunnels, long arcades, high bridges over rivers, or siphons across deep depressions was a last resort when difficult terrain could not be avoided. The former view, very popular among late nineteenth-century historians, that Roman engineers engaged in ostentatious displays of constructional expertise merely to impress, is not an opinion that is given much currency nowadays.

In addition to the constructional advantages of building aqueducts at or just below ground level, there were other benefits. Access for cleaning and repairs, for instance, was relatively simple and at intervals the conduit was equipped with *putei*, vertical access shafts. In modern times these *putei* have proved a great boon to archaeologists. Over the centuries workmen accumulated great piles of aqueduct debris – stones, gravel, silt and calcareous deposits – beside the *putei*, so that by locating them it has been possible to find the underground aqueduct. Other advantages of underground channels, or at least covered channels, were cleanliness, less evaporation and some insurance against water-stealing. Frontinus, however, makes it clear that the latter was always a serious problem. The idea that buried water-channels were a fully effective means of maintaining the supply to a city when under siege is not reasonable.

The prototype Roman aqueduct, the Aqua Appia, met Rome's demands for forty years until 272 B.C. when the Aqua Anio Vetus was built. Others followed at intervals until finally, when Frontinus became Rome's water commissioner in A.D. 97, there were nine aqueducts in all. Subsequently, three more were built and three of the older aqueducts were provided with additional supplies along new branches. For ease of reference, details of all Rome's aqueducts are set out in Figure 19.

From the details set out in the table a number of general observations can be made. The statistics relating to distances confirm the previous remarks about Roman engineers' reluctance to build bridges. The table shows that of something in excess of 300 miles of aqueduct, less than 40 miles were carried on arcades, and the true mileage of bridges was even less. Across the Campagna, both the Aqua Tepula and the Aqua Julia were carried on the Aqua Marcia's bridge, while the Aqua Anio Novus shared a bridge with the Aqua Claudia. This multiple use of two basic sets of arcades was feasible since a number of different aqueducts had their sources in the same general areas, and therefore were necessarily obliged to follow very similar routes to Rome. Hence the use of a single bridge to carry two or three water-channels presented itself as a logical expedient. Moreover, since such a recourse so obviously confounds adherents to the idea that the Romans indulged in structural displays for their own sake, it is surprising that the point has not been emphasized before. In reality only about 5% of Rome's aqueduct-mileage was carried on bridges.

The subsequent addition of a *specus* to the top of an existing aqueduct was technically not difficult. Concrete construction was used to graft both Tepula and Julia on to the top of Marcia and the outsides of both were faced with brick, in appearance a contrast to the heavy masonry construction below. The extension to the Aqua Claudia was of brick throughout with a concrete lining. In the long run, the considerable increase in superimposed weight proved structurally taxing for the arches

			Lengths (in miles)					
Name of Aqueduct	*Date*	*Total*	*Under-ground*	*On Surface*	*On Bridges*	*Cross-section (feet)*	*Delivery level (feet above) sea-level)*	*Source*
APPIA	312	11	10½	½	300ft	2¼×5½	65	Springs in the valley of the River Anio. Excellent quality.
ANIO VETUS	272	40	39¾	¼	—	3×7½	157	River Anio. Turbid water of inferior quality.
MARCIA	144–140	57	50	½	6½	5×8½	192	Springs in the Anio valley. Excellent quality.
TEPULA	126	11½	5¼	½	5¾	2½×3½	200	Volcanic springs in the Alban hills. Warm water at source.
JULIA	33	14¼	7¾	½	6	2×5	212	Springs in the Alban hills. Excellent cold water.
VIRGO	21	13	12	¼	¾	2×5¾	67	Springs in the Anio valley.
ALSIETINA	10	20½	20¼	—	¼	5¾×8½	55	Lake Alsietinus. Turbid and unpotable water used for a naumachia, street cleaning and irrigation.
CLAUDIA	38–52	43	33½	¾	8¾	3×6½	221	Springs near source of Marcia. Excellent quality.
ANIO NOVUS	38–52	54	45½	1½	7	4×9	230	River Anio above Subiaco. Turbid water, poor quality.
TRAJANA	109–117	37	37	—	—	4¼×7½	240	Springs near Lake Sabatinus Quality not known.
ALEXAN-DRINA	226	14	8	4½	1½	—	—	Springs near L. Regillus.
ALGENTIA	4th cent?	—	—	—	—	—	—	Springs on Mons Algidius.

Figure 19 Details of Rome's aqueducts.

below; opening of the joints between the masonry voussoirs was unavoidable and occasionally serious.

Another problem worsened these deformations which were sometimes sufficient to open cracks right through to the water-channel. The weight of two or three water-channels on one set of piers led, in time, to differential settlement which could easily overstrain the arches. The notion that temperature effects were especially serious in the very long arcades is not altogether acceptable. The strains produced per unit length are the same in both long and short sections. However, temperature effects should not be underestimated. Frontinus himself, in addition to remarking that 'damage is done either by the lawlessness of abutting proprietors, by age, by violent storms, or by defects in the original construction', also observes that an advantage of underground constructions is that they 'not being subjected to either heat or frost, are less liable to injury'.[4]

An interesting opinion of Frontinus is his view that some of the later works were the more defective in construction. That cracks in the arches of the aqueduct bridges manifested themselves quite soon is evident from the existing remains. Many of the Aqua Claudia's arches, for instance, have been heavily built-up with thick layers of brick, tile and mortar extending in some cases well down the piers. The aim may have been structural reinforcement, although Clemens Herschel once advanced the theory, very convincingly, that the intention was merely to plug leaks. He points out that repairs from within the overhead lengths of aqueduct would have been very difficult to execute and lengthy interruptions to the flow of water inevitable. So the repairs were tackled from outside.[5]

The table, Figure 19 on page 83, reveals what might conceivably be a significant sequence if one excludes Alsietina, the naumachia filler on the Tiber's eastern bank, Virgo, the supplier of two low-level baths, and Alexandrina. Starting with Appia, whose final delivery point was 65 feet above sea level, and progressing through to Trajana whose delivery point was 240 feet above sea level, it emerges that the delivery points of Anio Vetus, Marcia, Tepula, Julia, Claudia and Anio Novus are at progressively higher levels. So can it be reasonably concluded that an increasingly higher delivery point was the intention of each engineer who added a new aqueduct to Rome's system?

Estimates of Rome's size and population at different times, reached by various methods, suggest the following numbers: about 400,000 at the beginning of the first century B.C.; over half a million by about 50 B.C.; approaching one million in the first century A.D. and probably attaining that number in Frontinus' time; reaching a peak at something over one million in the middle of the second century A.D., and thereafter declining. This increase in population is consistent with a need for increasing supplies of water and the time span, say 100 B.C. to A.D. 150, broadly matches the period of major aqueduct building. Conceivably, as the population grew it became increasingly necessary to supply water to higher parts of the city. Thus one finds that Claudia was the first aqueduct capable of supplying water to the top of the Caelian hill and later on, by means of an extension, to the Palatine also.

Strenuous and elaborate efforts have been made in the past to calculate exactly how much water was supplied to ancient Rome and to show thereby that the *per capita* consumption exceeded that of any other city right down to modern times. Probably the latter is true, and there are certainly many large cities which even now are not supplied with so much potable water per head of population as was Rome. But quantification is another matter, and a very obscure one.

When Frontinus became Rome's water commissioner, the oldest supply, the Aqua Appia, was already over 400 years old. The aqueduct was as old to Frontinus as Drake's Leat at Plymouth is to us. Anio Vetus was well over 300 years old and Marcia nearly 250 years. Only Claudia and Anio Novus were, to Frontinus, of recent origin. So many years of use had taken its toll (and in the case of Claudia and Anio Novus, relatively recent construction had been offset by some shoddy workmanship). Leakage was a more or less permanent problem brought on by disintegrating linings, temperature and settlement cracks, and the intrusion of the roots of trees and bushes. Repairs bear witness to the continuous efforts made to plug the visible leaks and regular inspection by workmen entering via the *putei* no doubt helped to locate major cracks from within.

But in more than 300 miles of underground aqueduct the undetected leakage must have been substantial. Sometimes very considerable repairs had to be undertaken on the older aqueducts. Frontinus says[6] that when Marcia was being built, both Appia and Anio Vetus were renovated and that when Agrippa built Julia he also repaired Appia, Anio Vetus and Marcia. A variety of inscriptions record repairs to *all* the aqueducts in 5 B.C.; to Virgo in A.D. 31, 43 and 44; to Claudia and Anio Novus in A.D. 52; to Claudia in A.D. 71, after ten years of use and *nine of disuse*; to Marcia in A.D. 79 and Claudia and Anio Novus a year later; to Claudia again in A.D. 84; and to Marcia about the time of Frontinus' death. So any attempt to quantify rates of flow is bound to be modified by two considerations: at any one time it is unlikely that all the aqueducts were in commission, and those that were, leaked.

Then there was theft and fraud. Frontinus complains at some length about tampering with the size of outlet pipes, bribery, and direct theft of water *en route* to Rome. The attention he gave to these matters rather suggests they were serious misdemeanours widely practised. If so, the quantity of water reaching Rome might well have been significantly reduced below the theoretical maximum.

Calculations based on the physical dimensions of the water-channels are of little help. As the table (Figure 19, page 83) shows, all are at least 2 feet wide and, with one exception, at least 5 feet high. But this by no means represents the cross-sectional area of flow. Each *specus* was large enough to allow the entry and movement of workmen, which required a space much greater than that demanded by the flowing water. It is true that Frontinus lists the quantities which were *supposed* to flow in each of 'his' nine aqueducts. But as we shall see shortly, Frontinus' conception of flow measurement was seriously deficient.

Over the years, estimates of the quantity of water reaching Rome have varied from 20 to 400 million gallons per day. As guesses, any one is as good as another. As serious estimates, these wide-ranging figures, for the most part conscientiously evaluated, merely underline the impossibility of determining the capacity of Rome's water-supply system.

No other city in the Roman Empire was equipped with so large a system as Rome nor one built up over so long a period. Nevertheless, some of the provincial systems were sizeable undertakings and for the historian of engineering a few are of great interest and significance. The twin criteria of interest and significance must now be applied, for to list all Roman water-supply systems would require a book in itself. Suffice it to say that public water-supply was a feature of every part of the Empire in which Rome exerted her urbanizing influence. In some regions, particularly those which had previously fallen within the sphere of Greek influence, e.g. Pergamon, existing installations were sometimes expanded and 'modernized'. Generally, however, the concept, certainly on the Roman scale, was being introduced for the first time.

France is particularly rich in Roman aqueduct remains and cities such as Nîmes, Metz, Bordeaux (two aqueducts), Orange, Arles, Fréjus, Lyon (four aqueducts), Rodez, Paris (three aqueducts), Marseilles (three aqueducts), Aix-en-Provence (four aqueducts), Vienne (three aqueducts), Antibes (two aqueducts) and Vaison, can all boast remains of some description. By far the most impressive from a structural point of view is the famous Pont du Gard. Its three tiers of arches rise to a height of over 160 feet across the small River Gard, 20 kilometres north-east of Nîmes. The bridge is conventionally attributed to Agrippa, *c.* 20 B.C., and considering its great age the state of preservation is remarkable. Much can probably be attributed to the bridge's massive and precise construction in cut stone throughout. Apparently no mortar or iron clamps figure in any of the joints while a notable detail, undisturbed to this day, is a variety of slots and projections used by the Roman builders to support scaffolding and falsework. The Pont du Gard portrays very well the massive and laborious lengths to which Roman engineers were sometimes obliged to go merely to achieve a simple result.

Lyon, the Roman Lugdunum, was founded in 43 B.C. and became one of Gaul's most important administrative and military centres. To supply water to its steadily increasing population four aqueducts were constructed; opinions as to their dates vary. The first, the Mont d'Or aqueduct, is probably not as early as 43 B.C., a date often given. The Augustan period is more likely, and it may well have been Augustus who built the second supply, the Craponne aqueduct. Modern opinion ascribes the Brévenne aqueduct to Claudius, about A.D. 50, and the last, the Gier aqueduct, to Hadrian.[7]

The terrain on the western side of the Rhône, from which Lyon took its water, is difficult country for aqueduct building. All but the Craponne aqueduct followed the most tortuous routes in an effort to find a steady falling grade and the longest, Gier, had a total length on the ground of over 75 kilometres even though its source was only 40 kilometres away (very near modern Saint-Étienne). Lyon's engineers must have put considerable effort into locating and surveying the routes of the four aqueducts, but even so they were not in the final analysis able to avoid the problem of carrying the flow over some notably deep and very wide depressions. The result was nine inverted siphons: two on the Mont d'Or aqueduct, two on Craponne, one on Brévenne, and no less than four on Gier.

A number of Roman water-supply systems in Spain are a well known and important part of the story. The Tarragona aqueduct features a small two-tiered bridge which, but for the conduit's roof, is complete; Mérida's two aqueducts comprise not only some fine bridges but two large reservoirs as well; and most prominent of all Spain's Roman relics is the tall and slender bridge which brings water into Segovia. Recently restored and still in use, this mighty monument nearly 100 feet high dominates the Plaza de Azoguejo in Segovia. Perhaps because it is such a bold and slender structure, the Segovian aqueduct bridge used to be attributed to Hadrian's era but up-to-date opinion inclines to the view that it is Claudian. Certainly this is in accord with the style of construction: large granite blocks laid with smooth, unmortared joints and rough faced.

In the last few years the careful and thorough researches of Carlos Fernandez Casado and others have greatly extended our knowledge of Roman aqueducts in Iberia and the list of works is now very much longer.[8] Some of these aqueducts were not very large and their relatively small flow was only sufficient for towns of modest population. Here and there, however, more substantial supplies were procured, sometimes from considerable distances, and large-scale engineering was necessary. Toledo, for instance, even though it was not so very important a place in Roman Spain, drew a substantial water-supply from a large artificial storage reservoir 38 kilometres away to the south. By far the most taxing engineering problem was the crossing of the Tagus gorge. Fragments of masonry work in the gorge suggest to most modern opinion the one-time presence of an aqueduct bridge, a truly massive structure which would have needed to stand 300 feet high in order to carry an open conduit. Currently work is in hand to reconstruct in the Roman manner fragments of the bridge on the assumption that it was originally a three-tiered structure resembling the Pont du Gard.[9]

A puzzling aspect of the Toledo aqueduct bridge is its almost total disappearance. It is curious that such a huge bridge – if bridge there was – and probably the biggest of any the Romans built, should in fact have left the least evidence of its existence. We shall return tentatively to this point shortly.

The extent of Roman irrigation in North Africa, referred to in Chapter 1, is suggestive that the present-day climate is very much drier. Although less grandiose, Roman towns in Mauretania and Numidia were very numerous and important as military bases, as centres for the production of corn, timber, building stone and oil, and as ports. The water-supply systems of Roman Africa have by no means been so well researched as their European counterparts, but the indications are that similarly elaborate engineering was frequently resorted to. The spring-fed supply to Leptis Magna required the construction of two dams, one of them no less than 900 metres long. Roman Carthage can claim the distinction of being supplied along one of the longest of all Roman aqueducts, no less than sixty miles from springs at Zaghouan to the city. On the long aqueduct which supplied Chemtou were two bridges, and there was a very high bridge on the aqueduct at Tebessa. Structurally, the most notable aqueduct bridge is the one at Cherchel, a Hadrianic work standing 115 feet high.

One could add more examples, more details, more mileages of aqueduct and so on in order to emphasize the efforts of Roman society to provide people with adequate quantities of potable water free of charge. Bearing in mind an earlier point, namely that the basic technology had been worked out and applied in pre-Roman times, it is now worthwhile to investigate the extent to which Roman engineers improved and developed this basic technology themselves, what innovations, if any, were introduced by them and the influence which their example exerted on later societies.

An innovation which stands out is the concept of storing water for drinking behind dams. Storage in tanks and cisterns within towns was an old technique which the Romans copied but impoundment at source was something new. The three outstanding examples of storage dams are to be found in Spain, two at Mérida and one near Toledo.[10]

The Alcantarilla dam, which impounded water for use in Toledo, is probably the oldest of the trio. It was a composite structure employing a vertical water-face wall of concrete covered with masonry and backed along the air-face side with a heavy earth embankment. Ultimately this mode of construction was a failure. At some unknown date when the reservoir was empty the pressure of the earth pushed the front wall over and the dam was ruined. In the Prosperpina dam, the earlier of the two at Mérida, this defect was remedied. The water-face wall is buttressed at nine points where the dam is particularly high (about 12 metres) and the structure, despite centuries of neglect, is still sufficiently secure to remain in use.

Mérida's second water-supply dam, Cornalvo, was probably built in the second century A.D., as were the other two, but somewhat later. Its basically earthen construction is more sophisticated and its hydraulic apparatus for regulating the outflow and overflow is more elaborate and much better designed. With a total capacity of 10,000,000 cubic metres the Cornalvo reservoir was the largest in Spain until the late eighteenth century, and like its neighbour it is still in use.

The urge to store drinking water in large amounts is symptomatic of the Romans' commitment to large-scale water-supply. Wherever possible they turned of course to perennial springs and rivers, and yet when the need arose, in arid regions such as Spain, the job of building a reservoir to create a regular and reliable supply from a meagre and intermittent source was successfully tackled.

The Romans' use of hydraulic concrete, strong and quick setting even under water, enabled them to roof the Pantheon, build bridge spans of 100 feet or more, make solid and permanent bridge piers in rivers and carry out submarine harbour works. Concrete figured prominently in water-supply works as well. It was a basic structural material in all the big water-storage dams, was widely used in the construction and lining of water channels and allowed important developments in aqueduct bridge building.

Although precisely and beautifully constructed of cut masonry blocks, early bridges such as the Pont du Gard were structurally clumsy. Achieving great heights by stacking arched bridges on top of each other (Figure 20) was an effective technique but tedious and expensive. During the first century A.D. more elegant and structurally efficient shapes evolved, of which the Segovia and Tarragona bridges are good examples.[11] The bridge-on-a-bridge concept can be seen giving way to a more monolithic form, in which the piers are more nearly continuous from foundation to crest with a substantial degree of arched bracing between.

The application of concrete permitted even more refined and elegant bridge forms. In the Mérida aqueduct bridges (Plate 5) and the one at Cherchel, the tall slender piers are continuous from top to bottom; they consist of concrete cores faced with masonry and brick. The 'bridge-stacking' form has completely disappeared and longitudinal bracing was achieved principally from the water-channel and its supporting arches at the top of each pier. The lightweight jack arches at lower levels were probably needed to provide stiffness and stability during construction. Roman aqueduct bridges are a very instructive example of the process of structural evolution at work.

The use of the siphon by Roman engineers has been much underestimated and misjudged. In *A History of Western Technology* (1959) Friedrich Klemm adheres to the old view that the Romans' passion for ostentatious displays of constructional expertise favoured high bridges; R. J. Forbes has written: 'In Roman times the siphon was used only in some few cases, probably because of leakages and the relatively poor materials available for high pressures'; Thomas Ashby believed that the siphon was ignored because Roman engineers were unaware of the 'crushing-resistance of their hydraulic cement', an erroneous view because the strength of pressurized pipelines is in fact dependent on *tensile* strength.

Such judgements are not in accord with the evidence which establishes that Roman engineers frequently resorted to large siphons. If the famous group of nine siphons at Lyon be taken as typical, then lead was the material normally used to fabricate the tubes of siphons. Sufficient strength was developed by providing wall thicknesses of half an inch or more in pipes whose overall diameter was about ten inches.

In order to carry large quantities of water, amounts which would be commonplace in an open channel, an individual siphon would comprise up to a dozen parallel pipes. The quantity of lead required for a single

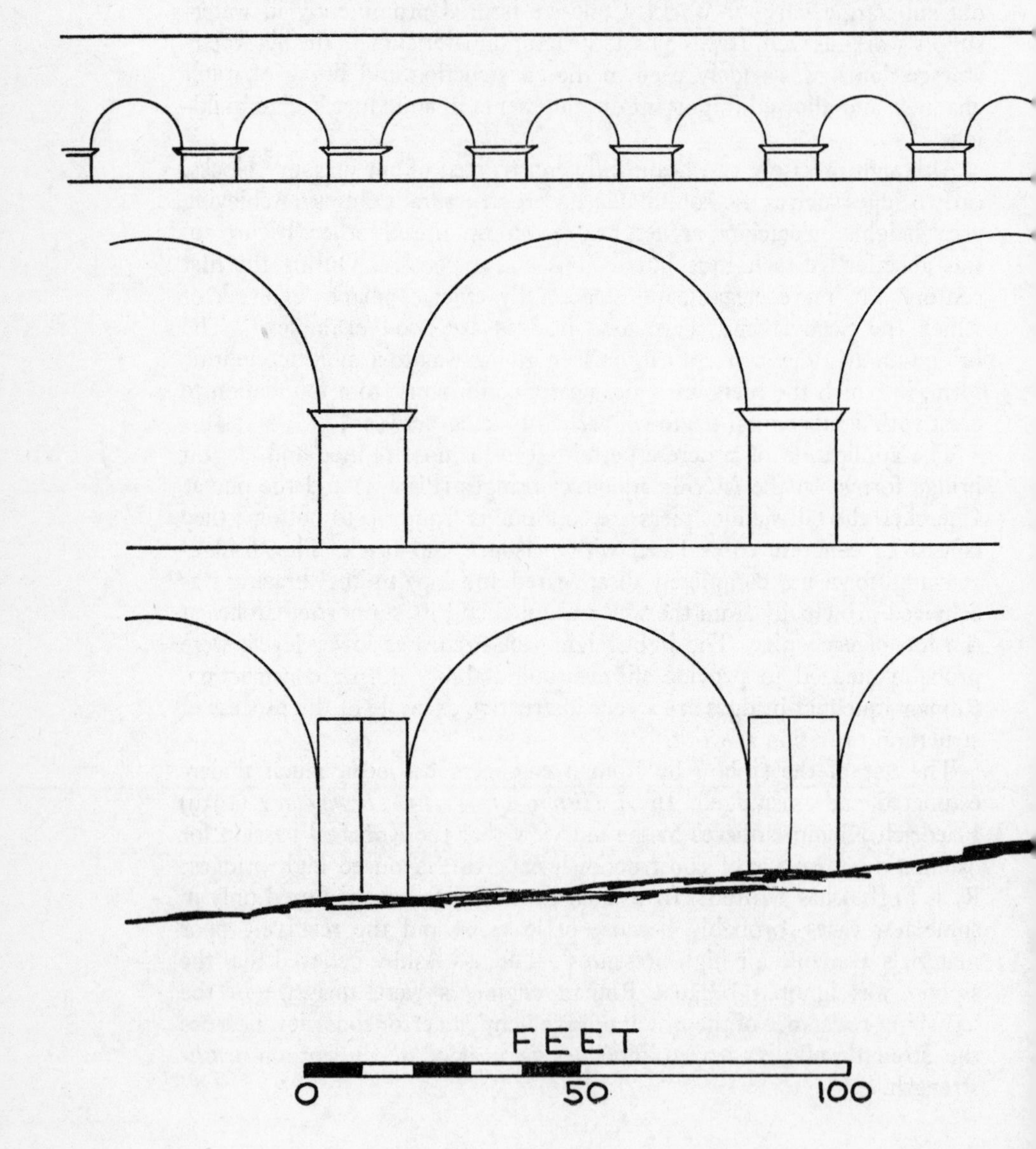

Figure 20 Elevations of three Roman aqueducts: *opposite* the Pont du Gard; *top*, Mérida; and *bottom*, Tarragona.

crossing of this type was prodigious. The large Beaunant siphon on the Gier aqueduct at Lyon is estimated to have contained at least 2,000 tons of metal. If Roman engineers were disinclined to build siphons at all, the problems of mining, processing, transporting, fabricating and laying such large weights of lead is the best explanation.

An immense commitment to pressurized lead pipelines is by far the most significant feature of Lyon's aqueducts and represents the most lavish use of siphons in antiquity. Today not a trace of the thousands of tons of lead which were used remains; doubtless it is now spread over the roofs of churches and buildings throughout Lyon and its environs. Nevertheless much has survived, especially on the Gier aqueduct, to fix the location, depth and length of the siphons. There are some header tanks complete with rows of holes into which the pipes were fitted, receiving tanks, traces of ramps down which the pipes were laid (Plate 6) and aqueduct bridges in the bottoms of the crossings along which the pipes were carried at their deepest points. These low-level bridges, sometimes very substantial structures as much as 60 feet high, were used to reduce the depth of the pipeline and hence contain the maximum pressure to manageable levels.

So far as can be seen at the moment, Roman engineers resorted to the use of inverted siphons when the topography was such that a bridge was judged to be impossibly high or long, or both. The length criterion is difficult to assess because many arcades are very much longer than the longest known siphons. The question of height, however, is more revealing, although the evidence is far from conclusive. The tallest Roman bridges – Alcantara, 160 feet; the Pont du Gard, 160 feet; Narni, 130 feet – are lower in overall height than the depth reached by any of the major siphons – for example; Beaunant (Lyon), 405 feet; Tourillons (Lyon), 380 feet; Alatri, 340 feet; or Rodez, 300 feet. In other words, whereas there is no example of a Roman bridge more than 160 feet high, and indeed only a handful over 100 feet high, the majority of siphons reached depths in excess of 200 feet, and not infrequently 300 feet.

Thus it seems plausible that a crossing depth in the range of 150–200 feet represented to Roman engineers an economic and practical limit to the construction of a bridge. For depths above 200 feet or so it was the inverted siphon which offered the more reasonable solution. On this basis the idea of a 300-foot high bridge at Toledo becomes untenable, and one is bound to postulate that in fact the Tagus was crossed by a siphon carried at its lowest point on a single-arched, one storey bridge. Such an explanation meets the objection raised earlier that practically no traces of a *very* large bridge survive at or near the site.

Two features of Roman siphons continue to be puzzling. It remains uncertain, for example, how siphons were started up from dry, a potentially damaging operation, and how they were drained if and when repairs were necessary. Conceivably, some sort of valve was fitted in the siphon's lowest level and it is perhaps to this device that Vitruvius alludes in Book VIII, Chapter VI of his *Ten Books* in a puzzling passage which has led a succession of modern writers to reach some strange conclusions and even stranger layouts for Roman siphons.[12] The notions that Vitruvius is discussing the problems of air-locking (dissolved air coming out of solution) or is describing some means to reduce the hydrostatic pressure in the siphon are both bogus.[13]

The second unanswered question is how Roman engineers matched the flow capacity of a pipeline to that of the open channels at either end. Insufficient flow in the siphon would, after all, cause the inlet tank at the end of the approaching channel to fill up and overflow, so wasting precious water. Roman engineers must have evolved some procedure for dealing with this particular feature of siphon design, although it is evident that they made no use of flow calculations. Quite apart from the intrinsically difficult problem of head losses due to friction in such long, narrow-bore pipes, an issue which was not fully resolved until the nineteenth century, the Romans were not even familiar with the method of computing quantities in an open channel.

When Frontinus discusses this most elementary question he reveals a startling ignorance of what is required.[14] While he realized that the same quantity of water ought to flow at any point in a given conduit – the continuity principle – he always calculates the quantity on the basis of cross-sectional area of flow alone. Nowhere does he give any consideration to velocity of flow. It is extraordinary that Roman engineers could have achieved so much in the way of useful water-supply using open channels and siphons without ever being familiar with the most basic calculation pertaining to the problem.

We have devoted some space to an outline of the more important Roman water-supply works and the engineering ideas and developments which went into their construction and operation. No apology is offered for this emphasis. Roman water-supply remains one of the great achievements of engineering in antiquity. The concept may not have been Roman any more than the basic technical components were – although significant hydraulic and structural advances eventually came about – but the vast and lavish scale on which the Romans operated is impressive by any standards. The notion of using civil engineering to supply society – all elements and not just certain privileged sections – with adequate clean water in tens of towns and cities was fundamentally new. Nothing like it had been attempted before and was not to be attempted again until the nineteenth century.

An essential problem facing European and American cities a century or so ago was water-supply on the Roman scale, and it is some tribute to Roman engineering that the solution was reached with fundamentally the same apparatus.

8

Fifteen Centuries of Neglect

THE USE OF the term 'neglect' is relative rather than absolute. For something like fifteen hundred years public water-supply was given precious little attention *by comparison* with the Roman period. Here and there, however, at different times, and mainly in the larger cities, water-supplies of various types were set up and some discussion of these will form the basic contents of this chapter.

Around A.D. 300 Roman rule in the West was in decline and a notable shift in power to the eastern half of the empire was under way. In A.D. 330 Constantine established his eastern capital at Constantinople, the old Greek Byzantium, and here a new centre of engineering grew up, with water-supply a prominent aspect.

Apparently the earliest component of Constantinople's water-supply was due to Hadrian, and this scheme probably formed the basis of the work of Valens in the fourth century and of Justinian in the sixth century. The oldest surviving relic of this early period is Valens' aqueduct bridge which stands in modern Istanbul within the boundaries of Constantine's original city. Justinian's contribution to Constantinople's water-supply was considerable during a reign which produced a notable burst of structural engineering generally. A number of old dams to the north of Istanbul, although probably not Byzantine in themselves, may well be Turkish reconstructions of something older, or they might be Turkish dams grafted on to older aqueducts.

Of more than three dozen bridges which carry Constantinople's aqueducts over depressions[1] most are believed to occupy the sites of Byzantine structures dating back to Justinian's reign or earlier. Just one bridge, about six miles north of Istanbul, is thought to have survived virtually unchanged from the sixth century. The contention that Justinian's engineers made special efforts to design against earthquakes, following severe damage to the Church of Santa Sophia soon after its completion, is difficult to test.

Apart from water-supply, the most notable engineering work in Constantinople was its fortifications which over the centuries were to be frequently and severely tested; and in order to survive a prolonged siege adequate and defensible bodies of stored water were necessary. Constantinople adopted the procedure which had long been a feature of cities such as Jerusalem, namely the construction within the city walls of a network of underground cisterns. Of more than forty investigated, all have been identified as being of Byzantine origin although Ottoman renovations and extensions are numerous. Most of these storage tanks, many of which still exist, were quite small, while of the larger ones two date from the sixth-century reign of Justinian. The roofs of both of these cisterns are carried on veritable forests of columns, in one case of marble topped by ornate Byzantine capitals, the appearance being reminiscent of 'a fine cathedral flooded out rather than that of a utilitarian reservoir'.

While the efforts of Roman and Byzantine engineers can be said to have equipped Constantinople with the basis of sound water-supply which lasted, the same was not true for cities in the West. While some Roman aqueduct systems did experience a degree of reconstruction and renewed use – Rome, for example, and Segovia – most fell into disrepair, the structures were destroyed or 'quarried' and the Roman concept vanished. Dark Age and medieval Europe with rare exceptions resorted to the most primitive of water-supply techniques, no better than those of the pre-Graeco-Roman world and not infrequently a good deal worse. Wells, springs and river-water were all that many communities could depend upon, and in most towns and cities such sources were inclined to be very unwholesome. In the more heavily populated towns, of northern Europe especially, the combination of people, the wet climate, bad and muddy roads, the movement of livestock and the total lack of proper drainage or adequate removal of sewage conspired virtually to guarantee the pollution of water sources, in particular springs and wells.[2]

That highly unsanitary conditions prevailed must have been manifest, and here and there one can find examples of legislation designed to control the indiscriminate dumping of refuse and sewage in streets and rivers.[3] The use of technology to effect a solution, however, was rare. Moslem Spain, in cities such as Cordoba and Valencia, made efforts to channel clean water from potable sources and in the twelfth and thirteenth centuries some not unsuccessful efforts to pipe water into Paris were instituted. Monastic communities frequently applied themselves to the task of piping or channelling spring water, but the resulting improvements were to their own advantage, not to that of society at large. Following the dissolution of the monasteries, however, a number of water-supplies to religious houses were commandeered for general use but with no great effect because of inappropriate location and inadequate quantities.

The sixteenth century, certainly in England, did at last witness the beginnings of a new effort to supply drinking water, a revival which in principle at least is reminiscent of Roman technology. In 1560, for example, Plymouth proposed to augment its limited reserves of well-water by diverting the flow of the River Meavy.[4] To some extent Sir Francis Drake was instrumental in pressing the plan, and work on the 17-mile-long conduit was eventually begun in 1589. Drake's Leat, as the aqueduct came to be known, was commissioned with due pomp in 1591.

Before 1600 London had made little attempt to pipe or channel water-supplies from clean sources outside the city.[5] The Thames continued to be the principal source, its tributaries such as the Fleet, the Walbrook and the Tyburn were used, and there were many wells. As the capital's population grew – and the increase was rapid by the standards of the time – these traditional sources became so increasingly inadequate in terms of both quality and quantity that plans to engineer new and more plentiful supplies began to materialize.

In 1609 Hugh Myddleton, citizen of London, Member of Parliament and goldsmith, privately sponsored a scheme to convey water from Ware in Hertfordshire to north London. Typically, vested interests immediately raised a variety of objections. Landowners along the proposed route visualized their property being swamped and divided by the presence of the aqueduct and they predicted danger to both men and animals with the risk of flooding compounding the disadvantages even further. Notwithstanding these obstacles, the construction of the 'New River' went ahead and Myddleton was able in 1613 to fulfil his promise 'to finish the same within fower yeares'.[6]

The New River was an open channel 10 feet wide and generally about 4 feet deep. From its source to a circular delivery pond at Islington was a distance of 38¾ miles. To maintain an even gradient, cuttings and embankments had to be formed and two small timber aqueducts lined with lead carried the channel over depressions at Edmonton and near the Islington terminus. There was access for repairs and maintenance along both sides, an unspecified number of wooden bridges crossed the channel and at a few places roads and tracks tunnelled underneath it (see Figure 21). Choosing and levelling the aqueduct's route proved to be the most demanding technical operation of all. In order to provide sufficient head below the delivery pond at Islington – it was some 82 feet above high river level – the fall of the aqueduct could be little more than 5 inches per mile. The required levelling was done by Edward Wright, a well-known mathematician and cartographer of the day, and the accuracy he achieved is suggestive that the art of levelling was, by his time, well developed. The efforts of Humphrey Bradley and his surveyors in the Fens just twenty years before point to the same conclusion.[7]

While the New River with the aid of many modifications, improvements and enlargements, was destined to last (as a private company until 1904 when the Metropolitan Water Board absorbed it), a contemporary scheme proved ultimately to be less enduring. However, the idea of pumping London's supplies from the Thames (and from other rivers in other cities)

Figure 21 Bush Hill embankment as rebuilt in 1682 to carry the New River over Salmon Brook near Edmonton. (From W. Matthew's *Hydraulia*)

is technologically of great interest and representative of a very distinct phase in the evolution of water-supply. To set the scene, a brief résumé of the history of the essential ingredient, the piston pump, will be helpful.

The two-cylinder piston pump of Ctesibius (see Figure 22) is known to us only through Vitruvius whose description in Book X, Chapter VIII is incontestable. It is clear that the Romans used piston pumps of this exact type but only in very small sizes. Of the handful of specimens which have been unearthed, at Silchester, Metz, Saint-Germain-en-Laye and Bolsena, for example, all measure only inches in both bore and stroke and apparently they were hand operated. Those which are not made of bronze comprise a wooden body with lead pistons. The Roman force-pump always worked submerged and never utilized a suction stage. The small size, the mode of construction and the manual operation of Roman pumps all suggest that their application was limited to minor tasks such as filling domestic cisterns.

We simply do not know if piston pumps were part of Dark Age and medieval European technology. Their appearance in various fifteenth-century technical manuscripts – such as those of Francesco di Giorgio

Martini, Taccola and Leonardo da Vinci – may merely be the first pictorial record of what was, in reality, a continuing tradition. If not, at least three possibilities are open to consideration: Europe inherited the idea from Islam whose engineers *were* familiar with piston pumps[8]; the discovery of Vitruvius's manuscript in 1408 prompted fifteenth-century renderings of the Ctesibian pump; or European technology brought forth the machine as part of current developments in which the introduction of the crank and connecting rod combination was important.

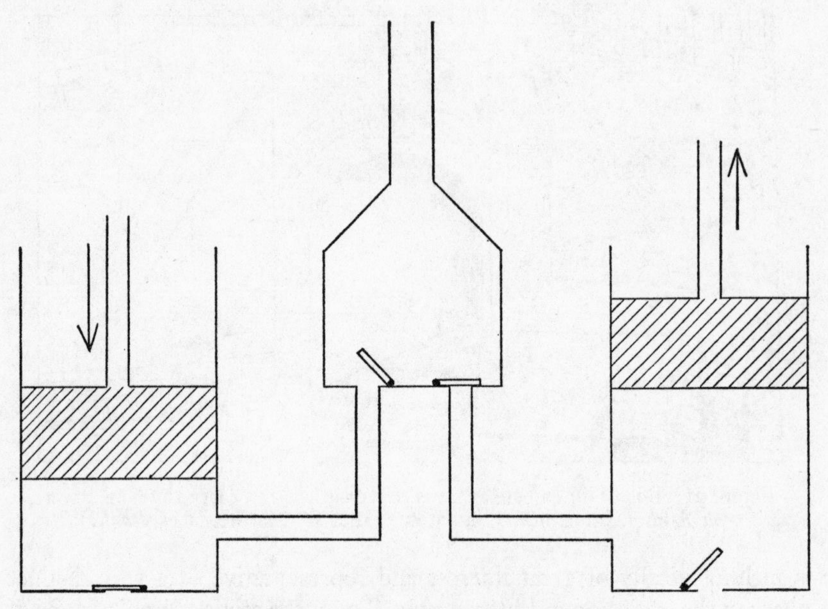

Figure 22 The two-cylinder piston pump of Ctesibius based on Vitruvius's description.

The straightforward force- or lift-pump has the great advantage that it will lift water to any height consistent with the piston, cylinder, valves and delivery pipe being able to withstand the hydrostatic pressure. It also has two notable drawbacks: the pumping mechanism is submerged and, worse, if the water-level falls, the cylinder will not fill. A solution is the use of a suction pipe on the inlet side of the pump. Not only does this allow the pump to be set above the water, it will also accommodate changes in the water's level. In theory the suction stage can be about 33 feet, the height to which atmospheric pressure will support a column of water, but in practice more than 25 feet is almost impossible to realize. The really interesting fact is that when the European piston pump makes its appearance in the fifteenth-century writings of Taccola (*c.* 1450) and Martini (*c.* 1475) the suction stage is already incorporated (see Figure 23). Indeed it predominates, in that the forced or lifted stage of the pump's action is little more than the stroke of the piston. The origins of this fundamentally important and ingenious modification to the basic piston pump are as obscure as the prior history of the piston pump itself.[9]

The introduction of the piston pump, with both suction and lifting stages, into mine drainage in Central Europe occurred at some date impossible to determine. However, the matter of fact descriptions and detailed drawings of Georgius Agricola suggest that German mines were thoroughly familiar with the piston pump as early as 1530. Of the seven varieties

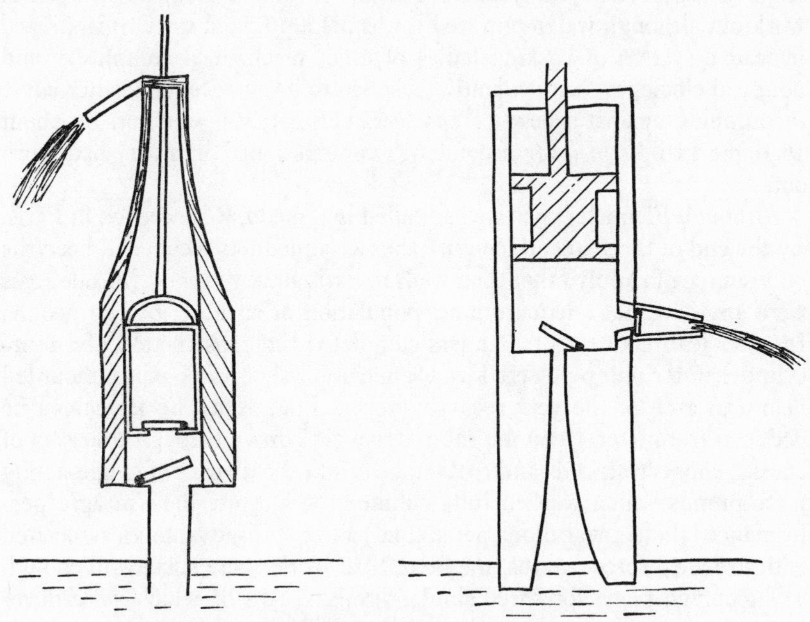

Figure 23 Two ideas for suction pumps as suggested by Francesco di Giorgio Martini in the fifteenth century.

of piston pump described by Agricola one, 'the most ingenious, durable and useful of all', he says was 'invented ten years ago'. Of another piston pump, water-powered, Agricola writes, 'this machine draws water through its pipes by discs out of a shaft more than one hundred feet deep'.

That the early part of the sixteenth century saw a rapidly widening use of the piston pump is borne out by the machine's first recorded application to water-supply. Again Germany is to the fore. The first evidence we have of pumped water-supply in Augsburg, Bremen and Danzig dates from the third quarter of the sixteenth century, but in reality these developments probably began much earlier. Augsburg was endeavouring to procure a water-supply as early as 1412 and the construction of towers, presumably components of some water-lifting machine, was subsequently undertaken.[10]

In 1526, when it was decided to improve the water-supply of the Alcazar in Toledo, a German engineer was employed. That he installed water-powered piston pumps is evident from the Chronicler of the Monastery de la Concepcion Francisca, who wrote: 'This device worked with great pistons, and the water hammered so furiously and was driven with such terrific force through the metal pipes that all the mains were fractured.' An attempt to rectify this initial failure was made by two Flemish engineers but without success. Not until 1569 was any sort of solution engineered. In that year Juanelo Turriano's famous *Artificio* was put to work but although water-powered it was not a pumped system; it utilized instead a system of rocking ladles of great mechanical complexity and doubtful efficiency, but evidently reckoned to be an acceptable alternative to pumping against a head of 250 feet. Ultimately it was not. By about 1640 the two *Artificios* (a second was commissioned in 1581) had worn out.[11]

Although Flemish expertise had failed in Toledo, it succeeded in Paris. By the end of the sixteenth century the two aqueducts which had been the only source of supply other than wells in Paris since the early Middle Ages were anything but adequate for a population of around 300,000 people. In 1608 Jean Lintlaer of Flanders completed the construction of a four-cylinder water pump powered by a single undershot water-wheel mounted below an arch of the very recently erected Pont Neuf. So far as can be deduced from later accounts, the water-wheel drove through four sets of cranks, connecting-rods and rocking beams to the quartet of single-acting force-pumps which worked fully submerged. Despite their meagre performance (about two gallons per second) and the disadvantages associated with drowned force-pumps, the Pont Neuf works were successful enough to encourage plans for three similar projects, two of which were never built (those of 1626 and 1656) and one which was. In 1670 an engineer called Joly provided the means to pump water from the Seine below the Pont Notre Dame.

Between 1663 and 1685 tremendous efforts were made to install a water-supply on a suitably lavish scale at Versailles.[12] Initially the works comprised an extensive system of reservoirs whose waters were moved from place to place by various groups of piston pumps powered either by horses or windmills. In 1678 a Liégeois called Arnold de Ville proposed to augment supplies even further by pumping water from the River Seine. By 1685 the massive and famous 'machine of Marly' was completed and operating, an achievement which in the main was due to yet another Fleming, Rennequin Sualem, a 'carpenter' of Liège.

The Marly machine derived its water-power from a low dam built across the Seine; fourteen undershot water-wheels, each one nearly 40 feet in diameter, drove three groups of force-pumps with suction inlets. Rennequin judged that a lift of 500 feet over a pumping distance of nearly one mile was impossible with a single pumped stage; the pressure would have been impossibly high and the effects of water hammer potentially damaging. So he provided 64 pumps at river level, 79 pumps at an elevation of 150 feet, and 78 pumps at a height of 325 feet. Other supplementary pumps totalled 38 in number, so the grand total was 259 pumps. The second and third stage banks of pumps were actuated by two rows of oscillating rods and levers, 'stangenkünste', an old idea for transmitting motion over a distance and the technique which Turriano had used at Toledo.

The Marly machine was the heftiest concentration of water-wheels, piston pumps and mechanical linkages which had been built up to that time. It became famous immediately and was often described during its 150 years of clanking and thumping existence. This lavish display of late seventeenth-century mechanical engineering produced at most 1 million gallons of water per day. By comparison, a work of civil engineering such as the New River, was able to realize close to 4 million gallons per day.

Water-powered water pumping was introduced on the Thames in 1582. The engineer in this case was a certain Peter Morris (or Morice), variously described as Dutch, Flemish and German. From a drawing dated 1635, we have a general idea of the layout of his pump which was installed under an arch of London Bridge where the ebb and flow of both tides and river turned the water-wheel. From the text which accompanies the picture a few details of the mechanism emerge but much remains obscure. Morris' pumps were all but destroyed in the Great Fire of 1666; subsequently they were got going again, but by 1700 they appear to have been in a poor state of repair.

At the beginning of the eighteenth century the repair of the London Bridge water-works and the construction of new pumps was undertaken by George Sorocold. By the time he had finished his improvements and additions (see Plate 7), the complete installations comprised four water-wheels driving between them a total of fifty-two piston pumps. Their combined performance is most difficult to determine. Henry Beighton's estimate, published in 1731,[13] was over 2½ million gallons per day raised through a height of 120 feet into the tanks which fed the network of distribution pipes.

Unfortunately, little is known of George Sorocold but evidently he was held in some regard as an 'erector of water-works', for he is variously referred to as a 'good engineer', 'ingenious' and 'the great English engineer'. Between 1692 and 1706, and in addition to his undertakings in London, Sorocold either built himself or was influential in arranging water-powered pumping plants in Derby, Norwich, Leeds, Bristol, Nottingham, Macclesfield, Wirksworth, Great Yarmouth, King's Lynn, Deal, Bridgnorth and Sheffield.[14] That one man could be involved in so many projects is perhaps indicative that around 1700 there was a general upsurge of concern over water-supply and while much that was projected was never realized the overall position was substantially improved.

During the middle years of the eighteenth century, London Bridge Works along with other lesser installations – such as York Buildings and Shadwell – appear to have functioned well enough but problems were at hand. The machinery itself, not well designed to endure decades of heavy vibration and high stresses in members and bearings, was slowly wearing out. And then in 1761 a serious blow fell. As part of a general renovation and remodelling of London Bridge two central arches were converted to a large single-span opening. It was a disastrous step. Quite apart from the damage which was done to the bridge's central foundation, the increased flow passage lowered the river and significantly impaired the performance of the water-wheels.

In 1767 an attempt to rescue the deteriorating situation was made by John Smeaton. At the southern end of the bridge he installed a 6-cylinder pump powered by a 32-foot wheel. These new works were basically the same as Sorocold's, but being the work of the finest engineer of the day were better designed and constructed. Smeaton however, unlike Sorocold, did decide to dispense with any apparatus for accommodating changes in water-level. This seems to have been a mistake; in dry summers and at low tides the performance of the so-called Borough Wheel was insufficient.

For a further half-century the London Bridge Waterworks continued pumping water to its ten thousand customers but virtually nothing was done to improve or extend the service. In 1822 came the decision to build a new London Bridge and the waterworks were demolished. It was about this time that other water-powered pumps on the Thames were replaced or dismantled, while on the Seine the Pont Neuf, Marly and Pont Notre Dame machines were given up in 1813, 1852 and 1858 respectively.

Whatever the shortcomings of water-powered pumps, in the long run they were not unsuccessful in meeting the immediate problem. Moreover, in order to use the supplies they provided, the basis of a network of distribution pipes was laid down and water companies, albeit small and private undertakings, were established. Ultimately, the existence of these companies, along with the experience and the capital which they had accumulated, was to prove valuable in promoting bigger and better enterprises.

As a factor in the evolution of mechanical engineering, water-powered pumping is of some significance. Although its component technologies – water-wheels, mechanical linkages, piston pumps – were for the most part evolved in response to problems in fields such as mine drainage and metallurgical work, their amalgamation to meet the needs of water-supply represents not only a new area of application but a notable development in terms of machine size as well. Evidently the reliability of water-powered pumping machines left much to be desired even in the better installations, but for all that a good deal was learned. By establishing which were the more efficient and robust mechanisms and which arrangements of water-wheels and pumps gave the better performance a useful advance was made. Engineers accumulated a valuable body of knowledge which made a useful contribution to power technology as the Industrial Revolution accelerated in the second half of the eighteenth century.

The fate of water-supplies pumped from rivers with water-wheels was ultimately sealed by two factors: the water obtained was insufficient in quantity and poor in quality. Undershot wheels, river-driven, would have needed to become increasingly large or increasingly numerous, or both, to keep pace with the growing demand which developed even in the eighteenth century; and in any case an economically more viable alternative was fast evolving. Thomas Newcomen's atmospheric pumping engine, introduced to drain mines, was not difficult to adapt to pumping from rivers where the required lifts were on the whole smaller, and there was no dead-weight of pump rods to lift.

The year 1726 witnessed the erection of the first two Newcomen engines for water-supply: one at York Buildings on the Thames, the other at Passy near Paris on the Seine. The York Buildings pump had a capacity, so it was claimed at the time, of around ¾ million gallons a day. In 1752 York Buildings acquired a second much bigger engine and thereafter others were installed on the Thames and the New River. It was in connection with improvements to the New River, in particular the effort to boost its flow, that John Smeaton in 1765 initiated his important search for ways to improve the Newcomen engine. As with his earlier research into water- and wind-power, Smeaton's scientifically designed and carefully conducted experiments yielded valuable results but an even more significant improvement was soon to exert far greater influence.

In 1776 the first Boulton and Watt separate condenser engine in London was built for a distillery at Bow. Two years later a similar engine was installed at Shadwell Waterworks, the first application in London to water-supply and apparently one of the earliest anywhere. In 1780 a condensing engine was exported to France to pump water from the Seine at Chaillot. In 1805 John Farey estimated that at least ten Watt engines were in use in the London area for water-supply.[15] Throughout the nineteenth century the steam engine held a dominant position as a driver of pumps wherever public supplies of water had to be lifted; water-power was ousted rapidly at the beginning of the century and electric pumps were coming in but slowly at the end. Basically it was James Watt's original beam engine and its various derivatives such as the Cornish and Woolf engines which were favoured, although other configurations were utilized especially for some of the bigger pumping stations built in the 1880s and 1890s.

Notwithstanding the steam engine's capacity to lift much greater quantities of water, it was ultimately no solution to the attempt to meet public water-supply requirements by pumping rivers; they were becoming much too polluted. Nineteenth-century conditions revealed this fact in no uncertain terms and the problem of supplying sufficient quantities of water was compounded by a need to ensure its quality.

From present day 'environmentalist' campaigns against the pollution of rivers it is not difficult to gain the impression that the problem is modern, having been brought on in the last twenty or so years by a sudden and unchecked demand for more power, a greater output of materials and so on. In fact, it is only certain aspects of water pollution – notably thermal pollution and some types of chemical pollution – which are of recent origin. In other respects the pollution of rivers has been a problem, and has been recognized as such, for centuries. A brief résumé of some of the evidence is in order.

It has already been pointed out that the River Tiber, especially in summer, must have become very inadequate as a source of public water-supply to ancient Rome. Emptying sewers into the river polluted the water to such an extent that alternative sources, in the form of distant springs, had to be utilized. In his *Quest for Pure Water*, M. N. Baker has detailed numerous examples of ancient techniques for purifying or at least clearing water which was judged to be unwholesome in its natural state.[16]

In the reign of Philippe-Auguste it is said that in Paris fish could be seen swimming in the Seine. Evidently this situation did not last and in 1404 Charles VI took steps to prohibit the dumping of filth in the Seine because it was 'an abomination to the sight and a pollution of the water'.[17] In London during the reign of Henry III it was recognized that the Thames was so foul as to be dangerous to health and a century later Edward III, in an effort to keep both rivers clean, forbade the use of the Thames and the Fleet as dumps for refuse and offal. Other parts of England in the fourteenth century were compelled to enact various laws against river pollution, and in 1388 these regulations were re-enforced by an Act of Parliament, conceivably the first time a government had ever taken such a step. Seemingly, however, this and later similar measures proved quite impossible to enforce. Rivers and streams continued to be wantonly contaminated and the widespread incidence of water-borne diseases along with regular outbreaks of violent epidemics is evidence that the polluted water was in general use for drinking. By the seventeenth century the problem was sufficiently serious to foster a variety of techniques and devices for cleaning dirty water. A good deal was written about it, but very little was done; and only rarely was any thought given to the question of preventing pollution at source, as opposed to purifying water already contaminated.

In the eighteenth century the theory and practice of water purification were frequently written about in England, Scotland, France, Germany and Italy, and in Britain a number of techniques and ideas were patented.

To some extent it was at last recognized that the contamination of drinking water could be either mechanical or chemical, and that these two features of the problem were essentially different. Practical measures for purifying water were limited to such expedients as boiling, adding various compounds which were thought to be beneficial, and filtering. The mechanical cleaning of water by means of sedimentation, a very old idea, and filtering through such substances as sand or charcoal, became established techniques; but the scale of operations was limited. Private filtering was popular among those wealthy enough to install it, while sheer necessity encouraged efforts to protect the health of armies and navies. Although the wholesale filtration of public water-supplies was occasionally advocated in the eighteenth century, and here and there schemes were prepared for particular towns and cities, nothing remotely in proportion to what was required was realized in practice.

Eighteenth-century concern over water purity was a response to a worsening social and technical problem. European cities generally were experiencing a substantial rise in their populations, while in England particularly the trend was the more pronounced because of expanding industrialization and the consequent growth of commerce. The population of London, something over one million at the time of the 1801 census, had perhaps as much as doubled in the preceding century, and much more explosive were the five- to ten-fold increases characteristic of places such as Birmingham, Manchester, Liverpool and Glasgow.

Heavy concentrations of people on the banks of the Thames, Mersey and Clyde, making do with sanitary arrangements and sewage systems which were inadequate even before the expansion of population began, guaranteed the unsuitability of river water for drinking. Nor was the position improved by the rapid growth of chemical and manufacturing industries who freely abstracted their requirements from rivers and irresponsibly poured their effluents back into them.

That the situation rapidly deteriorated in nineteenth-century London is evident from the response of the various Thames pumping companies. In 1821 the London Bridge Waterworks Company was close to admitting that it could no longer cope. Its prices were high, the delivery head available was difficult to increase and the water quality was in manifest decline. It was in fact admitted that pumping on the ebb tide yielded nothing but polluted water and the claim that standing in a cistern for twenty-four hours made it 'finer than any other water that could be produced' was feeble in the extreme.

By comparison, the New River Company continued to flourish. The 8 million gallons which it supplied daily were cleaner, cheaper and delivered at a substantially higher elevation than any water pumped from the Thames. It is significant that the New River alone provided some 65 per cent of London's water, and it is important to notice the success over more than two centuries of the concept of channelling good quality water from distant sources. Soon the technique was to be developed on a lavish scale.

That Paris also was facing a serious deficiency in the quantity and quality of its water is evident from a variety of sources and a notable technical innovation. In the middle of the eighteenth century Paris drew its water from three main sources: the Pont Neuf and Pont Notre Dame pumps on the Seine, and the Aqueduct of Arcueil bringing water from Rungis to the south. The population of Paris at the time was around half a million and consumption per head has been estimated at about 5 litres per day and falling. In 1760 when the Seine pumps, not for the first time, became defective, the position was so serious that the Académie des Sciences was moved to appoint a commission of investigation. Grandiose plans to channel very substantial quantities of water from rivers, first the Yvette and later the Bièvre, were initially thwarted by traditional obstacles – lack of money and the opposition of vested riparian interests – and finally abandoned when the French Revolution erupted. Steam pumping fared better. In 1778 at the instigation of the Administration des Eaux du Roi, two brothers, Jacques Constantin Périer and his brother Auguste Charles, were invited to install steam-powered water pumping on the Seine under the auspices of a privately financed body, the Compagnie des Eaux de Paris. At Chaillot and Gros-Caillou they successfully erected Watt pumping engines whose combined performance trebled the quantity of water supplied to Paris: the existing 2,500 cubic metres per day was raised to over 7,500 cubic metres, providing at least in theory about 14 litres per person per day.

As a private venture the Compagnie des Eaux de Paris expired in 1789 when it was taken over by the State. Under new management the Chaillot and Gros Caillou pumps remained in use until 1860 but by that date they had long since been replaced as the main source of supply in Paris. By 1800 the Seine had been recognized as an inadequate and unhealthy source and attention was focused once more on the notion of wholesale river diversion. In 1802 plans were approved to draw large quantities of water from the Ourcq, a tributary of the Marne to the east of Paris, and in 1809 the canalized supply was in flow. As a water-supply conduit, built to navigation canal dimensions, the Ourcq canal yielded a vast increase – about 60,000 cubic metres per day – in supplies to Paris, and this drastically reduced the significance of supplies pumped from the Seine.

A novel facet of the quest of Paris in the nineteenth century for more potable water was the development of artesian wells, the first of significant size being drilled early in the 1840s. Just when it was first appreciated that some aquifers are under such hydrostatic pressure that they will, if penetrated, flow to the surface of their own accord is a mystery. In the writings of the Arabic geographer al-Biruni there is a description of their mechanism and not long after, in 1126, it is recorded that artesian wells were bored at Lillers, in the Artois district – hence the name.

At an early date artesian wells were frequently bored in and around Modena in Northern Italy and the practice was so well established in the Middle Ages that the city's coat-of-arms depicted a pair of well-borer's augers. In his *Discours Admirables* of 1580 Bernard Palissy makes a fleeting reference to artesian wells, and the use of artesian wells in Northern Italy prompted two accounts in the seventeenth century. Giovanni Cassini, as a young man an engineer in papal service, was Professor of Astronomy at the University of Bologna from 1650 to 1671 during which time he 'observed that in many places of the territory of Modena and Bologna in Italy they make themselves wells of springing water'. That the whole subject was currently of great interest is evident from the work of Bernardino Ramazzini who late in the seventeenth century was Professor of Medicine in the University of Modena. In 1691 he published *De Fontium Mutinensium admiranda scaturigine tractatus physio-hydrostaticus* wherein are described the 'Wonderful Springs of Modena'.[18]

Late eighteenth-century records reveal a number of artesian bores in England. In 1794, for instance, a 260-foot well was sunk at Norland House near Holland Park; a flow of 46 gallons per minute was delivered 18 inches above surface level. In 1785 Erasmus Darwin contributed an article to the *Philosophical Transactions* entitled 'Of an Artificial Spring of Water' in which he describes his efforts to construct an artesian well near his home on the River Derwent in Derby. Darwin's effort is of interest in that he accurately anticipated the well's success by observing the local geology and hydrology.[19]

Prior to the nineteenth century, artesian wells were often constructed in Europe at those few favoured locations where the phenomenon was known to occur. Scientific exploration for new sites, Darwin to the contrary, was not undertaken. Artesian water was exceedingly small in quantity and highly localized in its use. In no sense did artesian water contribute to any comprehensive water-supply system; such a development was, however, to come in the nineteenth century.

9

Nineteenth-Century Revolution

IN 1829 AN ENGINEER called Murot successfully drilled artesian wells at Epinay and Suresnes near Paris. A year later a Parisian engineer by the name of Degousée bored an artesian well at Tours to the considerable depth, for the time, of 120 metres. The scientist François Arago, appointed director of the Paris observatory in 1830, was impressed by these projects and intrigued by the possibility of obtaining much bigger flows. It was Arago's belief that the water-bearing Albian greensand extended under Paris and would contribute usefully to the city's water-supply needs if it could be reached.

In 1833, at Grenelle, M. Murot began to bore for artesian water with the most primitive equipment, no more than an 18-metre gin and a pair of treadwheels powered by up to twelve men. The work was destined to go on for years. When a depth of 500 metres had been reached in October 1839, the well was still bone dry. Much criticism and even ridicule was levelled at Murot's techniques and Arago's ideas; neither, however, was deterred. Eventually, after more delays due to fractured tools (Figure 24) and with the power needed to turn the drill increasing substantially, the greensand was finally struck in February 1841 at a depth of 548 metres. The result was well worth the effort. Water spouted to a height of 33 metres above ground level at a rate of 4,000 cubic metres per day, a very substantial boost to the supplies of Paris. Large were the crowds which came to Grenelle to gaze at the jet which was soon to be enclosed in a tall and ornate tower.

The Grenelle artesian well was not only significant for being a substantial source of good quality water. François Arago's enthusiasm for the project in the first place, and his subsequent resolve to press on in spite of the immense difficulties, were sustained by the conviction that artesian water *ought* to be obtainable in such geological conditions, imperfectly though these were understood. The successful outcome was a convincing practical demonstration of the value of applying geological knowledge to the technology of water-supply. The work at Grenelle is an important landmark in the development of hydrogeology and it prompted a number of similar schemes and further developments and enquiries.[1]

In Great Britain, the earliest of all nineteenth-century efforts to procure water-supplies adequate in both volume and quality were made in Scot-

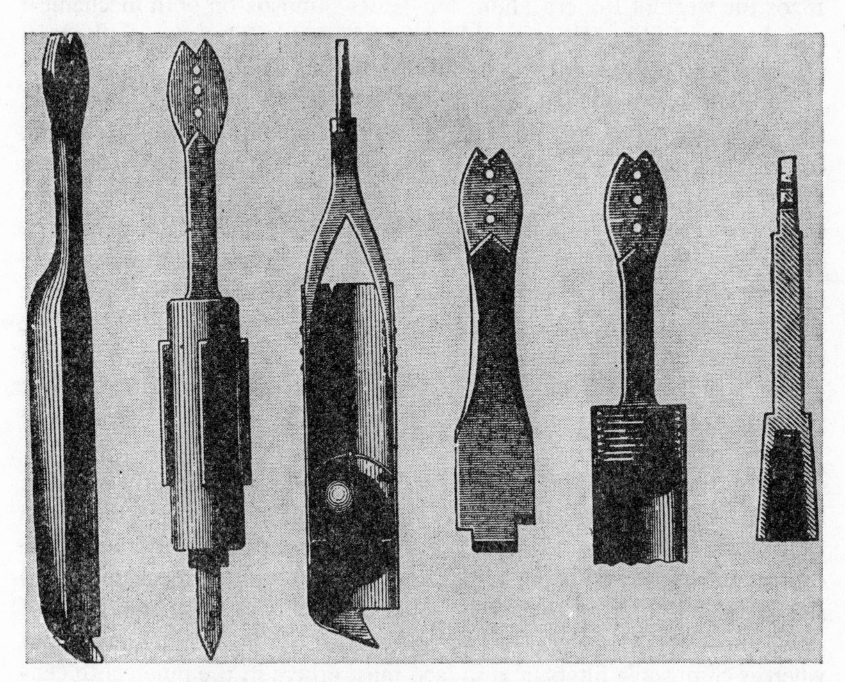

Figure 24 A selection of drill heads used to bore the Grenelle artesian well between 1833 and 1841.

land. Glasgow, with one of the fastest growing populations of any city in Britain, drew its water from installations on both sides of the Clyde. The bulk of the supply came from the northern bank and was pumped, by means of six steam engines, across the river in a cast iron pipe, an ingenious flexible device suggested by James Watt. It was laid in 1810, the hinged joints being provided to facilitate laying the pipe in the first place and subsequently to accommodate shifts in the river bed. This apparently complicated arrangement must have worked well in practice because eventually 28-inch and then 36-inch mains of the same hinged type were added to the original 15-inch pipe.

As early as 1804 Paisley's water-supply was filtered, the first public supply in the world to be so treated. According to some surviving but incomplete correspondence between Thomas Telford and the firm of Boulton and Watt, plans to filter Glasgow's water were first prepared in 1806 but the performance of these and numerous subsequent filtration units was at best feeble and at worst negligible. Of Glasgow's several pioneer water companies only one, the Glasgow Water Works Co., successfully filtered water in the early days using a gallery installed in 1810. Nevertheless, the idea of water filtration was beginning to attract some attention at the time, in England as well as in Scotland, and by the late 1820s the work of Robert Thom and James Simpson on both mechanical and slow sand filtration was yielding useful results; Thom was influential principally in the Glasgow area and Simpson in London.

Through his work on water-mills in Rothesay, Robert Thom was already experienced in problems of water impoundment and aqueduct construction when he undertook in 1824 to alleviate the scarcity of water at Greenock, a town which had not been well served by its first water-supply reservoir at Whinhill.[2] Using a combination of artificial reservoirs, open channels, automatic sluices and regulating basins, Thom was able to procure a daily supply of ten million gallons, so much in excess of domestic requirements that the bulk was used to power Greenock's mills. When the scheme was brought into use in 1827 its principal storage reservoir, Loch Thom, was the largest artificially formed body of water in Britain, and it was to retain this distinction until the completion of Lakes Vyrnwy and Thirlmere in the 1890s.

Two features of Thom's Greenock scheme were significantly original: his lavish use of ingenious and very reliable automatic flow regulators to control the outflow from reservoirs, to release compensation water and to control the supply to water-mills; and his provision of mechanical and slow sand filtration on the public branch of the supply. Thom's use of slow sand-filters was a year or two earlier than Simpson's in London but whereas Simpson's filters, and indeed most others in the nineteenth century, were cleaned by manually removing the clogged layers at the top of the filter, Thom resorted to the then very novel step of flow-reversal, a procedure also utilized in self-cleaning filters which he installed at Paisley in 1838 and at Ayr a year or two later. In 1829 Robert Thom reported on proposals for new water-works at Edinburgh. In the Scottish capital, however, Thom's influence was overshadowed by that of other engineers.

In the seventeenth century Edinburgh had already begun to extract water from the Pentland Hills to the south. In 1755 the supply was increased by piping the flow of springs at Bonaly, first of all with wooden conduits and later on with cast iron ones. In 1778 John Smeaton advocated further development of the Pentland Hills' springs and subsequently both Thomas Telford and John Rennie corroborated his views; following a water shortage in Edinburgh in 1810, Telford in the following year suggested how the water of Crawley Spring might be added to the system. Immediately the question of compensation water was raised.

In the latter decades of the eighteenth century canal building had frequently interfered with the régime of rivers to an extent which threatened the downstream interests of water-mill operators and landowners. Canal engineers such as Telford and Rennie had been involved in many a struggle over water-rights and Rennie was responsible for the design of a variety of devices which would automatically meter compensation water from reservoirs, locks and canals. During the nineteenth century the question of compensation water assumed great importance; the relatively small-scale difficulties of the canal building era were enormously magnified when late nineteenth-century demands for domestic and industrial water grew so dramatically.

To overcome the problem at Edinburgh in 1811 Telford proposed to build, on Glencorse Burn below Crawley Spring, a special reservoir from which as much water as was taken from the stream by diverting the spring could be released downstream in the dry season. The plan was not put into effect at the time although subsequently, especially in the north of England, compensation water reservoirs were not infrequently adopted in an effort to placate anxious and sometimes angry mill-owners. Where it was possible compensation water reservoirs impounding a pre-arranged portion of the run-off from a given catchment area were a sound idea, because they placed the management of compensation completely in the hands of the interested parties. Generally speaking, though, it was more usual, if reservoirs were built at all, to use their storage first and foremost in the interests of water-supply (or power) with compensation water being released as required and according to whatever arrangements had been agreed.

In 1818 another and more serious shortage of water in Edinburgh at last prompted some action. James Jardine, an Edinburgh civil engineer, was appointed to carry out the scheme which had been outlined by Telford and Rennie. Jardine seems to have coped manfully with some difficult practical problems, notably in connection with Glencorse dam. Both Rennie and Telford were absolutely insistent that unless the dam was securely founded on rock, perfect safety could not be ensured. Their standards were high and Jardine was only able to meet their requirements after months of laborious excavation.

The Glencorse dam and Crawley aqueduct were put into service in 1823. Following the drought of 1836, which laid the reservoir dry, new springs were tapped and more dams were built so that by the middle of the century there were six reservoirs in all. Edinburgh's daily supply was now nearly 30 gallons per head and all of it good quality spring water. When a Parliamentary committee, appointed in 1828, was making its report on the water-supply of London, it remarked of the Edinburgh works:

> Under the able direction of the late Mr Rennie, Mr Telford and Mr Jardine, and at an expense of only £175,000, the most magnificent works of the kind in Great Britain have been completed. The water is excellent; and the quantity to an inhabitant is nineteen gallons per day; and no less than 280,000 gallons are daily permitted to run waste. In real utility, they rival the boastful aqueducts of ancient Rome, and are the admiration of scientific strangers.

The use of the adjective 'boastful' can be dismissed as a typically misguided and unreasonable nineteenth-century view of Roman engineering, but in a general sense the parallel is valid. Before 1850, Scotland had gone as far as any country in resurrecting the Roman concept of water-supply. The reason was very likely a combination of several factors. There was in Scotland a long-standing tradition of interest in medical science and a greater concern for hygiene. Given the determination to procure supplies of potable water, there were adequate quantities available and they were not too distant from important centres of population such as Glasgow, Edinburgh, Greenock, Paisley, Ayr and so on. And Scotland was a breeder of capable engineers; not just the famous names such as Telford, Rennie and Watt, but also men like Robert Thom and James Jardine, less well known but highly competent practitioners all the same.

In England, the majority of water-supply works of nineteenth-century origin are to be found in the north. The upland water resources of the Pennines and the Lake District, which had stimulated the growth of many industrial centres in the first place, were increasingly harnessed to meet the needs of water-supply as urbanization concentrated and spread. By about 1850 there were a dozen or so large water-supply dams already built in England, and towns such as Bolton and Manchester, Sheffield and Huddersfield were already active in attempting to solve the water-supply problems brought on by urban congestion on the one hand, and massive and unchecked river pollution on the other. In contrast, London continued to do virtually nothing about what was the most appalling drinking water problem in Britain, and even possibly in Europe; radical changes, however, were not far off.

The 1828 Royal Commission on London's water-supply and James Simpson's pioneer filtration installation at the Chelsea Waterworks Company's plant beside the Thames, were both a response to a most singular event of 1827. On 15 March of that year, a man called John Wright published a thick and fiercely worded pamphlet entitled, for short, *The Dolphin or Grand Junction Nuisance*. Complete with an engraving (Figure 25) showing that the dolphin, or water-intake, of the Grand Junction Water Works Co. was located very close to the outfalls of several sewers, including one from the Chelsea Fever Hospital, John Wright's publication described the state of the Thames and the water abstracted from it in no uncertain terms. For instance, there was no mincing of words when he claimed that the river was 'charged with the contents of more than 130 public common sewers, the drainings from dung hills and laystalls, the refuse of hospitals, slaughter-houses, colour, lead, gas and soap works, drug mills and manufactories, and with all sorts of decomposed animal and vegetable substitutes'.

Wright's pamphlet, undoubtedly sensational and probably exaggerated, was at least instrumental in persuading Parliament to appoint a Royal Commission. Its report of 1828 was feeble and at its most censorious could

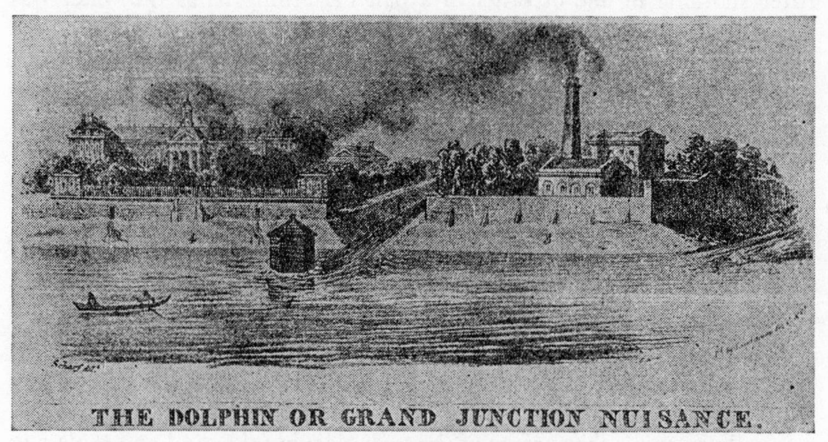

Figure 25 John Wright's graphic comment on the unhealthy location of the Grand Junction Water Works intake.

do no more than observe 'that many of the complaints respecting the quality of the water are well founded, and that it ought to be derived from other sources than those now resorted to, and guarded by such restrictions as shall at all times ensure its cleanliness and purity'. At least the commissioners saw fit to reject one particularly absurd piece of evidence with which they were presented, namely the claim that so long as sewage was emptied into the river near the banks it would not mix with the clean water flowing in mid-stream.

One of the three members of the 1828 Commission was Thomas Telford, and between 1831 and 1833 he served again, this time as the sole commissioner in a new and much more searching enquiry.[3] The result was plans for a comprehensive and integrated scheme excluding abstraction from the Thames altogether; but the estimated cost, over £1 million, was prohibitive at the time and nothing was done.

James Simpson's first interest in water filtration took the form of some small-scale experiments which he initiated in 1825 or 1826.[4] Following the publication of *The Dolphin* and the public and parliamentary outcry which it provoked, the directors of the Chelsea Water Works Co., anxious no doubt to place themselves beyond the strictures of Wright's pamphlet, urged Simpson to undertake a tour of existing filtration beds at the same time as his first working-scale filter at Chelsea was being constructed. During the summer of 1827 Simpson visited installations, mainly for the cleaning of *industrial* water, in Manchester, Glasgow and other northern towns.

This tour convinced him of the value of water filtration and clarified his ideas of filter design. Although small in scale, his prototype filter of 1827 set the pattern for all the later and larger permanent units which he built. After standing in one or other of a pair of settling tanks the water was introduced, gently, to the top of a 4-foot-thick downward flow filter. The filter media were graded from top to bottom; fine to coarse sand for a depth of 2 feet and then fine to coarse gravel for a further 2 feet. In the base of the filter, open-jointed brick tunnels carried away the cleaned water.

By his own admission, Simpson's first efforts were a failure, something which, if nothing else, convinced him of the value of preliminary experiments. Gradually, however, he improved the composition and layout of the filter bed and it operated very successfully for some months between 1827 and 1828. The turbidity of raw Thames water was substantially reduced and maintenance of the filter was not difficult. Most of the sediment was filtered off by the top ½ inch of sand – and none penetrated more than 3 inches – so that the filter's performance was easily sustained by manually scraping the surface of the bed, cleaning the sand and replacing it. Occasionally in his Chelsea and other filters Simpson supplemented hand cleaning with reverse flow washing.

Once the trial filter had been proved on a large scale, the Chelsea Waterworks Co. proceeded to build a larger permanent unit (Figure 26); it was brought into use early in 1829. Considering the usefulness of Simpson's filters and the manifest seriousness of the problem they were designed to tackle, it is extraordinary that for many years the Chelsea installations were the sole examples. Not until 1846 did Simpson advise on similar filters at York, followed by others for the Lambeth Water Co. in 1851, for the Chelsea Co. at Surbiton in 1856 and finally at Aberdeen in 1864. The reasons for the slow adoption of sand filters were essentially three-fold. There was in the first place a marked scepticism in many quarters that filtration was worthwhile. The removal of mechanical impurities, however desirable as a means of improving the sight, taste and smell of drinking water, did not, it was felt, guarantee any improvement in people's health. It was to be some years before it was appreciated that filtration was bacteriologically as well as physically beneficial. In view of the first reservation the other two are not altogether unexpected: problems of cost and space, especially the latter. James Simpson's Chelsea filter of 1829 had a capacity of no more than three million gallons per day and

PLATE I Geared water-raising wheels of the traditional type are still in use. This Spanish example shows the crudely-fashioned wooden driving wheel which is turned by an animal pulling on the beam at the top left.

PLATE 2 This water-powered noria at Murcia in southern Spain is of modern iron construction but the supporting structure is much older, certainly medieval and possibly Moslem.

PLATE 3 The Tibi dam as illustrated in the beautiful book by A. J. Cavanilles, *Observaciones sobre la Historia Natural, Geografía, Agricultura, Poblacíon y Frutos del Reyno de Valencia*, Madrid, 2 vols., 1795 and 1797.

PLATE 4 On the left is the modern Denver Sluice across the River Ouse above King's Lynn; to the right is the New Bedford River or Hundred Foot Drain which was cut in 1651.

PLATE 5 The impressive remains of the Los Milagros aqueduct bridge at Mérida. The piers are of concrete faced with brick and tile; only the springings of the jack arches remain; nothing is left of the water-channel except its supporting arches.

PLATE 6 Remnants of the Beaunant siphon on the Gier aqueduct at Lyon. The covered cistern at the highest point received water from an open channel and then directed the flow into a series of lead pipes laid along the ramp sloping away to the right.

PLATE 7 Henry Beighton's drawing of George Sorocold's water-works at London Bridge shows the installations as they were in 1731.

PLATE 8 The ornate and extraordinary outlet tower which draws water from Lake Vyrnwy for the supply of Liverpool.

PLATE 9 This scale model of the type of horizontal water-wheel used in Cordoba's flour mills has a diameter of 8 inches. It was probably used as a mill-wright's pattern for the construction of the full-sized article, some five feet in diameter.

PLATE 10 An outward flow pressureless turbine of the Girard type.

PLATE 11 A small low-powered Pelton wheel of the 1890s. The double buckets with jet splitters are still of a crude form prior to William Doble's radical improvement.

PLATE 12 A Kaplan propeller turbine for the Birsfelden plant in Switzerland. Under a head of 9.5 metres this 320 ton unit develops 22.1 MW. (Courtesy of Workshop Photo ESCHER WYSS)

PLATE 13 The runner from a Dériaz turbine installed at the Valdecanas hydro-electric station in Spain. Under a head of 245 feet and at 150 r.p.m. the turbine produces 108,500 h.p. The adjustable blades are here in the open position. (Courtesy of GEC Turbine-Generators Ltd.)

occupied in conjunction with its two settling basins a total area of about two acres.

So notwithstanding the potential value of Simpson's slow sand filtration and despite the recommendations of the Royal Commission (1828) and Telford's Report (1833), little was done to improve London's water-supply. The campaigning, however, went on, although no longer at the hysterical level adopted by the author of *The Dolphin*. A principal influence was Edwin Chadwick, who ceaselessly advocated that civil engineering, by providing a plentiful supply of piped, clean water and by arranging adequate sewage removal and treatment, held the key to the improvement of

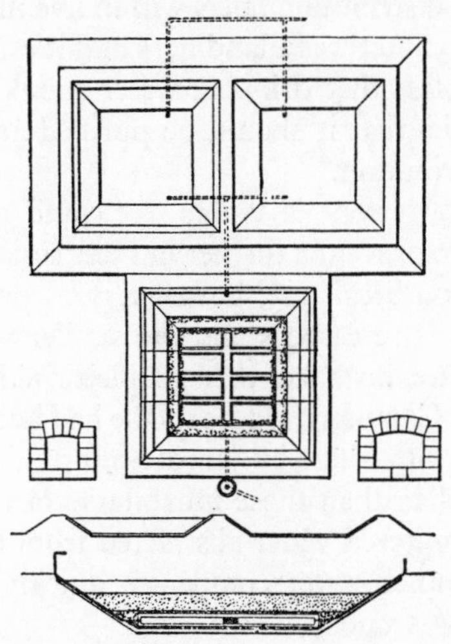

Figure 26 The arrangement of James Simpson's original slow sand-filter of 1829. (From *Proc. Instn. Mech. Engnrs.*, 1916)

urban conditions and the health of the poor. A more immediate influence, on occasions, was the effect of the stench which emanated from the Thames. In the 1850s both Parliament and the Law Courts were periodically reminded of the gravity of the Thames' pollution by 'an insupportable invasion of the noxious vapour' of such pungency that not even soaking the blinds over open windows with chloride of lime could prevent the adjournment of meetings. *Punch* lampooned the situation on more than one occasion, notably in 1855 following the publication by *The Times* of a graphic letter from Michael Faraday.[5] Ultimately, however, the one factor which above all others generated some action was the most deadly: cholera.

In the nineteenth century there were four serious outbreaks in England of Asiatic cholera, a disease which had never been experienced here before. In total they killed something in excess of 100,000 people. The first epidemic of 1831–2 was of itself sufficient to suggest that cholera was a water-borne disease but convincing proof was impossible to assemble. During the outbreak in the years 1848 and 1849 real proof still remained elusive but the circumstantial evidence was overwhelming. Among others, Dr John Snow was convinced of the connection between cholera and the drinking of contaminated water and he expresssed this view in *On the Mode of Communication of Cholera* of 1849. The hypothesis gained sufficiently wide acceptance to contribute in 1852 to the passing of the Metropolis Water Act. In essence the Act demanded three things: for drinking purposes Thames water could not be abstracted from the tidal stretch below Teddington Weir; it became illegal not to filter the river's water; and all reservoirs and distribution tanks within five miles of St Paul's had to be covered. The 1852 Act was a landmark and it established the essence of three basic principles: that drinking water should be taken from the purest source available, that it should be purified and that it should be protected during distribution.

Unfortunately, the Metropolis Water Act could not be implemented with sufficient speed to prevent a further but less serious cholera epidemic in 1853–4. The new outbreak did, however, provide more confirmatory statistics for Dr Snow's researches and the similar views of Dr William Budd. It was found, for instance, that people drinking the water of the Lambeth Waterworks Company, whose intake had been moved to Thames Ditton and equipped with filters by James Simpson in 1852, were far less prone to contract cholera than those inhabitants of the capital who continued to consume unfiltered water abstracted from the lower reaches of the river. London's other water companies quickly complied with the requirements of the Metropolis Water Act.

The fact that the purity of water-supply and the incidence of disease were related, fostered many enquiries, developments and changes in the second half of the nineteenth century, first of all in England and then in continental Europe and North America. The pioneer and uncertain work of John Snow and William Budd led to much more thorough research, culminating in the germ theory of disease. The work in the 1870s of Pasteur, Koch, Eberth and others founded the modern science of bacteriology, which not only revolutionized medical science but fundamentally altered the methods and, more important, the purposes of water filtration. Bacterial purity began to be given as much, if not more, consideration than mechanical and chemical purity.

In Great Britain two Royal Commissions (1865 and 1868) investigated river pollution and water-supply more thoroughly than ever before and the result was the passing of two famous acts of Parliament: the Public Health Act (1875) and the Rivers Pollution Prevention Act (1876). Within no more than a quarter of a century a thoroughly disorganized and irresponsible state of affairs had been replaced by standards that were high, comparatively, and regulations so stringent that they were not at all easy to meet. The water-supply engineer found himself confronted with some demanding problems.

In London serious thought was given to the very distinct possibility that the Thames could never again be used as a source of drinking water, as Telford had advocated years before. And generally speaking it is true of the late nineteenth century that abstracting river water within cities, for several hundred years a normal procedure because it was cheap and easy, was finally abandoned. The requirements of purity and the problems of quantity could not easily be met. London of course was in a difficult position. The Thames was by far the most plentiful source that was economically accessible. More than once the idea of bringing water to London from as far afield as Wales and Cumberland was entertained and as many times the idea was rejected as being hopelessly expensive, quite apart from which other big cities were casting covetous eyes on these northerly and westerly gathering grounds. Eventually it was decided that given a high degree of purification and with supplementary supplies being drawn from the River Lea and underground sources in Kent, the Thames could continue as London's main source of water. It was recognized, though, that this conclusion was inconsistent with the continued existence of eight separate private water undertakings. After decades of financial and political wrangling, the Metropolitan Water Board was formed in 1902.

The unavoidable shift away from rivers as sources of water-supply stimulated developments in the second half of the nineteenth century whose origins can be traced to the first: the exploitation of underground resources on the one hand, and upland resources on the other.[6]

Underground water sources are found throughout England but those which are substantial in volume are concentrated mainly in the east, the south and parts of the midlands. The Trent Waterworks at Nottingham appears to have been one of the earliest undertakings attempting to provide a public supply from groundwater sources. The works were built in 1830 by Thomas Hawksley, an engineer who was substantially involved in the construction of many water-supply systems and became a determined advocate of the social benefits they produced.

In the nineteenth century a few areas in Britain – the North and East Ridings of Yorkshire, for instance – came to depend almost entirely on underground water sources, while in other places – the London area, for example – they were developed as a vital supplement to other supplies. Overall, however, underground resources could only ever meet a portion of the total demand and for the most part it was the upland resources which were exploited most vigorously, some would say ruthlessly.

Impounding reservoirs in distant catchment areas feeding long-distance aqueducts, the technique to which parts of Scotland and northern England began to resort before 1850, proliferated after that date. Britain is not by world standards a country of big dams or large reservoirs – far from it – but it is an interesting thought that in terms of numbers Britain, with over 400 large dams, ranks high in the world list and leads in Europe. Possibly the reason for this lies in Britain's being so far ahead in water-supply legislation and public health provisions at an early date. A large number of public authorities, in the main of fairly small size, were obliged between them to construct a multitude of relatively small-scale water-supply works. Of most significance to our theme, however, is a handful of very large undertakings which were laid out in the last quarter of the nineteenth century.

The 1820s, especially the speculative period of 1825, was a decade during which joint-stock companies established several new water-works in Lancashire and Yorkshire towns, notably at Sheffield, Ashton, Oldham, Bolton, Preston and Manchester. They contributed a valuable increase in water-supplies at the time but were rapidly outgrown by a fast rising population and expansion of industry, especially in the Manchester area. In 1844 John Frederic La Trobe Bateman, who along with James Simpson and Thomas Hawksley pioneered modern water-supply engineering, was consulted by the Manchester and Salford Waterworks Company on 'the best means of obtaining an ample additional supply of water for Manchester and its neighbourhood'. At thirty-four years of age, Bateman was already a skilled and experienced hydraulic engineer; he had advised on major reservoir works in Ireland, in association with William Fairbairn, at Glossop and at Bolton. He was, moreover, one of the first, along with Robert Thom, to appreciate and advocate the absolute necessity of long-term rainfall measurements if upland water resources were to be properly developed. To this end Bateman began to establish rain gauges of his own and to record the readings from other people's whenever and wherever the opportunity arose.

Bateman's report on possible new sources opens in true Victorian style: 'Within ten or twelve miles of Manchester, and six or seven miles from the Gorton reservoirs, there is a tract of mountain land abounding with springs of the purest quality.' He was referring to a 3,675-acre catchment area east of Staleybridge which embraced numerous Pennine watercourses. A series of compensation water demands proved so impossible to meet, however, that Bateman's original plan had to be abandoned. Instead, the Manchester Corporation, having dealt successfully with various manoeuvrings designed to thwart their ambitions, plumped for the wholesale exploitation of the River Etherow and its tributaries along six miles of the Longdendale valley (see Figure 27).

The details of the design and construction of the Longdendale scheme were fully set out in 1884 by J. F. Bateman in his *History and Description of the Manchester Waterworks*, one of the classics of water-supply literature. In its final form the Longdendale scheme was as elaborate and comprehensive as any so far built in Europe or North America. It comprised five

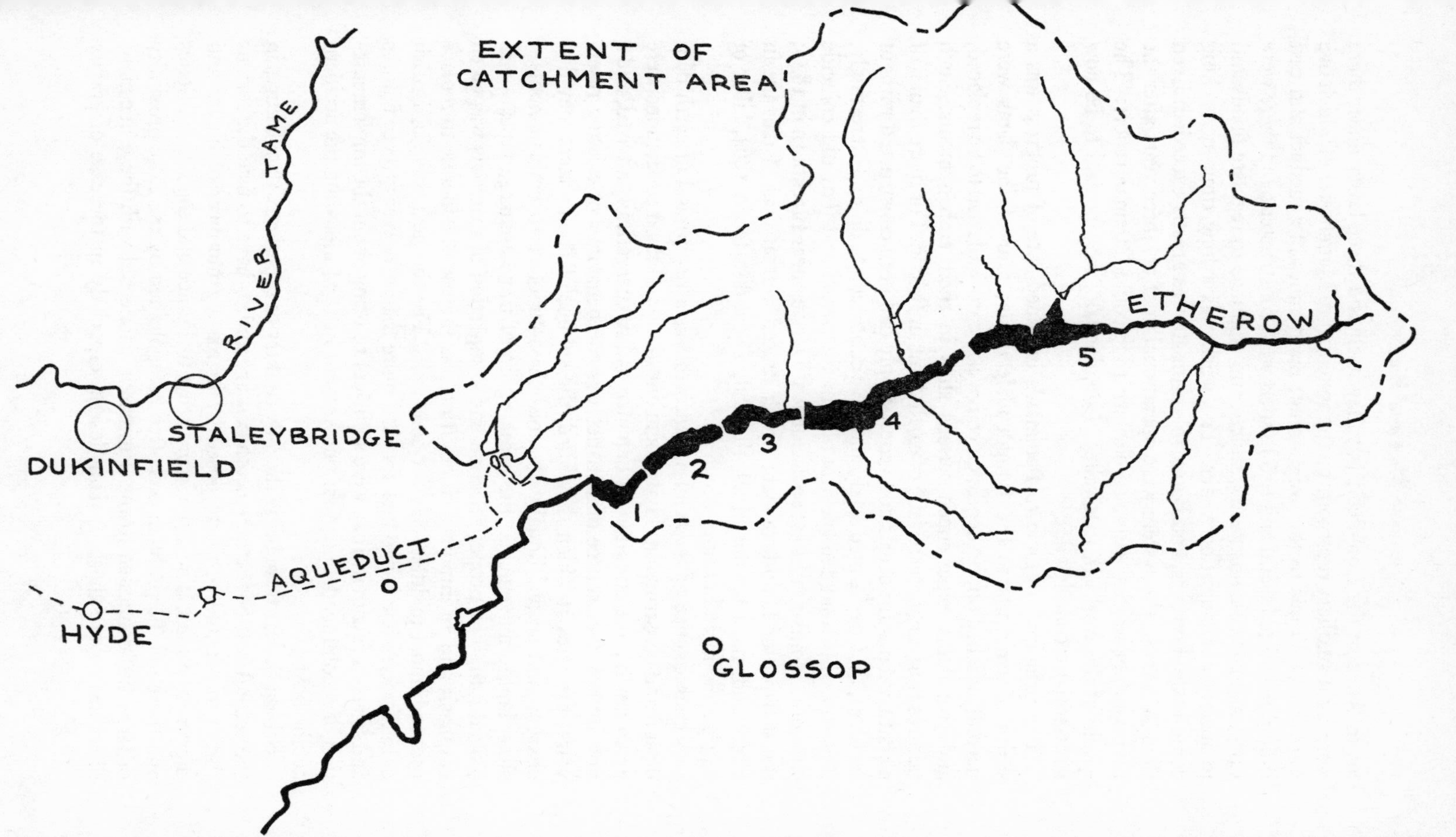

Figure 27 J. F. Bateman's five major reservoirs across the River Etherow in the Longdendale valley were: 1, Bottoms; 2, Vale House; 3, Rhodes Wood; 4, Torside; and 5, Woodhead.

major dams in the Longdendale valley itself and three lesser ones; there were eight auxiliary reservoirs on the aqueduct to Manchester of which five were late additions to the programme; many miles of aqueduct and cast iron pipe had to be laid and two tunnels were constructed; and there were various auxiliary dams, weirs, sluices and channels to regulate floods and to supply compensation water. The work was a long time in reaching completion, however, and Bateman's estimate was greatly exceeded. Even though parts of the system were operational in 1851, three years after the first contract was let, only in February 1877 could Bateman report: 'The whole of the works in that valley [Longdendale] may be said to be now substantially completed.'

The principal obstacle to Bateman's intended rate of progress was a series of accidents to the Longdendale dams; four of the dams were partially washed out by floods, defective clay core-walls had to be replaced, and two dams experienced serious slips in their embankments. Such failures were uncharacteristic of earth dams in Britain. For more than half a century dams for canals and water-supply had been constructed without incident, they performed safely and have continued to do so. Interestingly, however, the Longdendale troubles in the 1850s and '60s did coincide with other similar failures: an accident at Darwen near Blackburn in 1848; the collapse of Holmfirth dam which caused a great deal of damage in 1852; and, most serious of all, the Dale Dyke disaster in 1864, killer of 244 people in and around Sheffield.[7]

A consequence of these misadventures was a reappraisal of earth dam design and construction. Hydrologic questions and flood predictions were given much more attention, as Bateman had so energetically advised, and a not unimportant source of statistics was the behaviour of existing reservoirs: the rates at which they filled, the volume of flood waters they discharged and so on. Equally the misbehaviour of earth embankments – slips, bodily movements, the breaking up of iron discharge pipes – was looked into. The proper selection and compaction of materials was given emphasis and an important innovation was the use of concrete to replace the traditional puddled-clay for core-walls. For the next wave of British dam disasters one has to look forward more than fifty years; three failures and nearly a fourth in 1925 were sufficient to bring about the implementation of recommendations first put forward in 1864 following the accident at Sheffield.

Britain was far from being the only country to experience dam failures in the second half of the nineteenth century and to have to face the serious threat to life and property posed by increasing numbers of bigger and bigger reservoirs. The same issues arose in France and the United States and there too the problem was often compounded by the close proximity of large and congested towns. Intimately connected with these questions of security were those of dam design, especially in the case of gravity dams; the design of earth and arch dams remained essentially empirical until the twentieth century.

When the French government decided in 1858 to construct a flood-control and water-supply reservoir near Saint-Étienne, precedent and a mistrust of earth construction for high dams led to the choice of a curved masonry gravity dam for the purpose. Moreover, economic factors dictated that the dam be rationally designed using stability and stress analysis techniques pioneered by M. de Sazilly in 1853 and improved by Emile Delocre a few years later (see Note 9).

The design, construction and successful operation of Saint-Étienne's Furens dam revolutionized masonry dam-building. Engineers rapidly assimilated the new approach and although it was not applied rigorously in every case, which sometimes had serious consequences, dam designers were enabled for the first time to tackle questions of economy and safety with the confidence engendered by rationally formulated analysis capable of general application. What for centuries had been essentially an art was given a scientific dimension for the first time. Dam design quickly caught on in Great Britain.

The Longdendale scheme may have promised a plentiful supply for Manchester in the 1840s but it was far from able to cope thirty years later when the population had doubled. By the 1870s every substantial source within easy reach of Manchester had been appropriated by neighbouring towns and so it was necessary to go much further afield, to Cumberland, first of all to Lake Thirlmere and later on to Haweswater, both a good eighty miles from Manchester.

The Thirlmere scheme is significant for a number of reasons.[8] The aqueduct of over 95 miles in length was, with the exception of the contemporary Vyrnwy pipeline, an order of magnitude longer than anything built before, and nearly half of it was laid in cast iron siphons reaching as much as 9½ miles in length and, in one case, over 400 feet in depth. A Roman engineer would have been impressed. The storage reservoir, Lake Thirlmere, was a basically natural formation, its capacity increased by means of a 50-foot-high gravity dam, one of the earliest English examples of the type and rationally designed according to the precepts first evolved by de Sazilly and Delocre and subsequently improved by Professor Rankine.[9]

The engineering of the Thirlmere scheme presented no special difficulties and was completed in just under nine years. However a variety of non-technical issues, very much a portent of things to come, presented major obstacles and they were not overcome either easily or cheaply. The list of objectors was long and comprised powerful interests including railway and canal companies, town corporations, industrial factions, landowners and even the ratepayers of Manchester. Those who could not be won over had to be bought off or granted concessions which in some cases added appreciably to the cost of the work. Parliamentary expenses were exceptionally heavy in view of the determined opposition.

The Thirlmere scheme raised another issue; for the first time a major civil engineering work was bitterly opposed an aesthetic grounds. Threatening the beauty and solitude of the Lake District, more highly prized for its scenery than any other part of Britain, incensed feelings throughout the country. There was total conviction in the minds of many that Thirlmere's dam and aqueduct would cause ugly and permanent disfigurement of the landscape. In reality these fears had less foundation than the more sensitive critics imagined and in the final analysis the scenic disasters they conjured up never materialized. Nevertheless, important principles were raised and

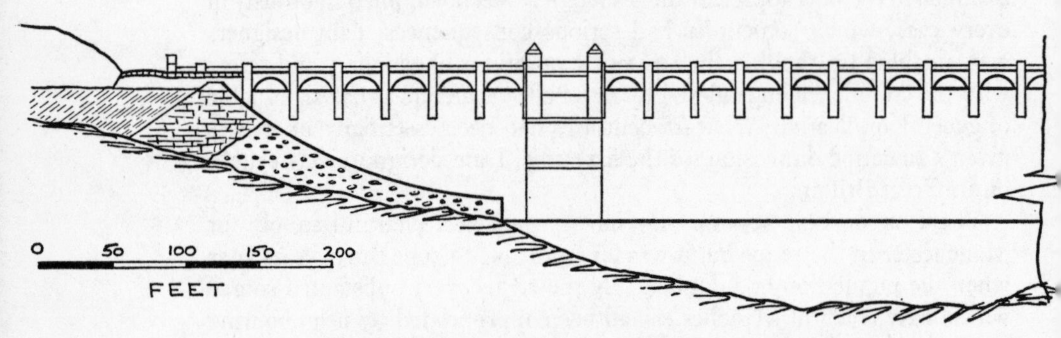

Figure 28 A half-elevation of the Lake Vyrnwy dam.

the 'sentimentalists', as protagonists of the project deridingly called them, propounded their views with great vigour and conviction in the late 1870s. Anyone who believes that environmental conservation is a modern cult should examine the sample of letters and articles reproduced in Appendix V of Sir John Harwood's *History and Description of the Thirlmere Water Scheme* of 1895.

For a brief period in 1875 it was envisaged that the water resources of the Lakes might be developed by the corporation of Manchester acting jointly with that of Liverpool. The notion was quickly put aside, however, and Liverpool turned instead to North Wales. In the valley of the Vyrnwy, a tributary of the River Severn, it was decided to form the largest reservoir in Europe and from it to pipe water sixty-five miles to Liverpool through a sequence of tunnels, open-channels and cast iron siphons, as in the Thirlmere scheme. Indeed the Vyrnwy project was exactly like that of Thirlmere in all its essentials and a close contemporary (1881–92) of it.[10]

The Vyrnwy scheme's most outstanding feature is its dam (see Figure 28), higher and longer than Thirlmere's, an overflow dam throughout its length and unmistakably Victorian in its massive appearance and ornate finish. George Deacon, the dam's designer, while convinced of the essential correctness of the continental approach to dam design was less certain that its procedures were sufficiently rigorous, and so he prepared for Vyrnwy a profile with a more than ample safety factor.[11] For added security the dam was constructed in cyclopean masonry to exceedingly high standards and it was equipped with an elaborate network of drainage tunnels as well. Of the many nineteenth-century water-supply works which survive in Great Britain, the Vyrnwy dam and its ancillary neo-Gothic outlet tower (Plate 8) are the perfect monument to Victorian civic pride, conservative engineering and cheap labour. The fact that a whole village lies drowned in the depths of the reservoir is almost completely forgotten today, nor does it seem to have been much of an issue at the time.

The Vyrnwy scheme was the first to impound water on Welsh soil for use in an English city. The exercise was immediately repeated. Birmingham, with a population of over half a million, seven times what it was around 1800, had depended, typically, on local rivers and wells for a water-supply throughout the century. The inadequacy of these sources was acknowledged in 1890 and had in fact been manifest for years before. For a greatly increased supply Birmingham chose gathering grounds surrounding the valleys of the Elan and Claerwen west of Rhayader in central Wales. From an engineering standpoint the work followed precisely the layout of the Thirlmere and Vyrnwy schemes with one notable exception: a total of six storage reservoirs, three on each river, were envisaged rather than one. In the event, only four of these were built as part of the original works and they came into service in 1904. Not until 1952, when the Claerwen dam was completed, was that river brought fully into the system.

The water-supply works of Manchester, Liverpool and Birmingham, all brought into service just before or just after 1900, mark the culmination of the nineteenth-century revolution. The concepts first expressed at Greenock in 1796 were utilized more and more as the century advanced. Between 1850 and 1900 over 140 water-supply dams were built in the British Isles, more than were constructed between 1900 and 1950. While three of England's biggest and fastest growing cities very naturally produced by far the most elaborate and expensive projects, the pattern was repeated on a descending scale in virtually every county. The Roman concept had not only been resurrected, it had been surpassed in every respect.

10

Other Countries, Another Century

A VERY DISTINCTIVE FEATURE of long distance water-supply in Britain was an almost total reliance on cast iron pipes for aqueducts, frequently containing very high pressures when deep valleys had to be crossed. Aqueduct bridges comparable in size to the high viaducts used on canals and railways were never built, and in a country well versed in the techniques of manufacturing cast iron cheaply this was logical. In continental Europe, however, and in the United States, a few big aqueduct bridges were constructed.

Both France and Portugal had examples before 1800. Incorporated in the water-supply system of Montpellier, built between 1753 and 1766 by Henri Pitot (best known for his invention of the Pitot tube for measuring water and air velocities), is the massive and beautiful bridge of St Clément, strikingly reminiscent of the Pont du Gard and not surprisingly, since it was Pitot who cleverly and tastefully grafted a road bridge on to the side of the Roman structure in 1745. Lisbon's famous Alcantará aqueduct bridge, part of the Aqueducto das Aquas Livres, is not only famous for its great height (225 feet) and length (3,000 feet) but more significantly because it survived the Lisbon earthquake of 1755. The grey marbled columns topped by pointed arches were seen to sway and tremble but nothing fell.

The highest of all aqueduct bridges was built between 1839 and 1846 on the 51-mile-long aqueduct laid down to supply Marseilles with water from the River Durance. Again there is a resemblance to the Pont du Gard. The Roquefavour bridge (Figure 29) is three-tiered to a height of nearly 100 metres, the 15-metre spans of the two lower tiers being replaced by a multitude of 5-metre spans in the arcade at the top. The Canal of Marseilles attracted the interest of many engineers outside France and was deemed sufficiently important to warrant a paper in the *Proceedings of the Institution of Civil Engineers* (Vol. 14, p. 190).

The classic American example of an aqueduct bridge is the famous High Bridge (Figure 30) which carries the old Croton Aqueduct across the Harlem river. Indeed it is a very early example in the United States of a large bridge of monumental character for any purpose. The Croton scheme, built between 1837 and 1842, enabled New York City to be furnished with adequate water-supplies for the first time, although not before some protracted legal and administrative wranglings had been resolved and a variety of difficult engineering problems overcome. Nor was the supply achieved cheaply; the Croton aqueduct cost the city almost $13-million

Figure 29 Elevation of the Roquefavour bridge of 1846, 270 feet in height and the highest of all aqueduct bridges. (From *Min. Proc. Instn. Civ. Engrs.*, vol. XIV)

although one must bear in mind how large a work it was, larger in fact than any scheme of a comparable type in any other city of the United States or Europe. Croton's engineer, John B. Jervis, not only had to cope with the construction of High Bridge (and some lesser aqueduct bridges), he was also obliged to dam the Croton River, a difficult job made worse by floods which washed out the nearly completed structure; there were 42 miles of aqueduct to be laid, for the most part a uniformly graded masonry conduit 8½ feet by 7½ feet in cross-section; and at the New York end receiving reservoirs and a distribution system had to be built. The claim that this 'was one of the most notable public works of the nineteenth century' is not unreasonable, certainly so far as the United States is concerned.[1]

The Croton scheme and another for which Jervis was the engineer, the Cochituate project at Boston (1846–8), are representative of a rapid transition in American water-supply which was taking place around the middle of the century.[2] Of the handful of organized water-supply schemes that had been previously arranged, all were small scale, all but one was privately owned and, for the most part, all were notably inefficient. Springs, wells and direct pumping from rivers, without storage, were the only sources. In 1652 Boston had instituted a very feeble supply by conveying spring water

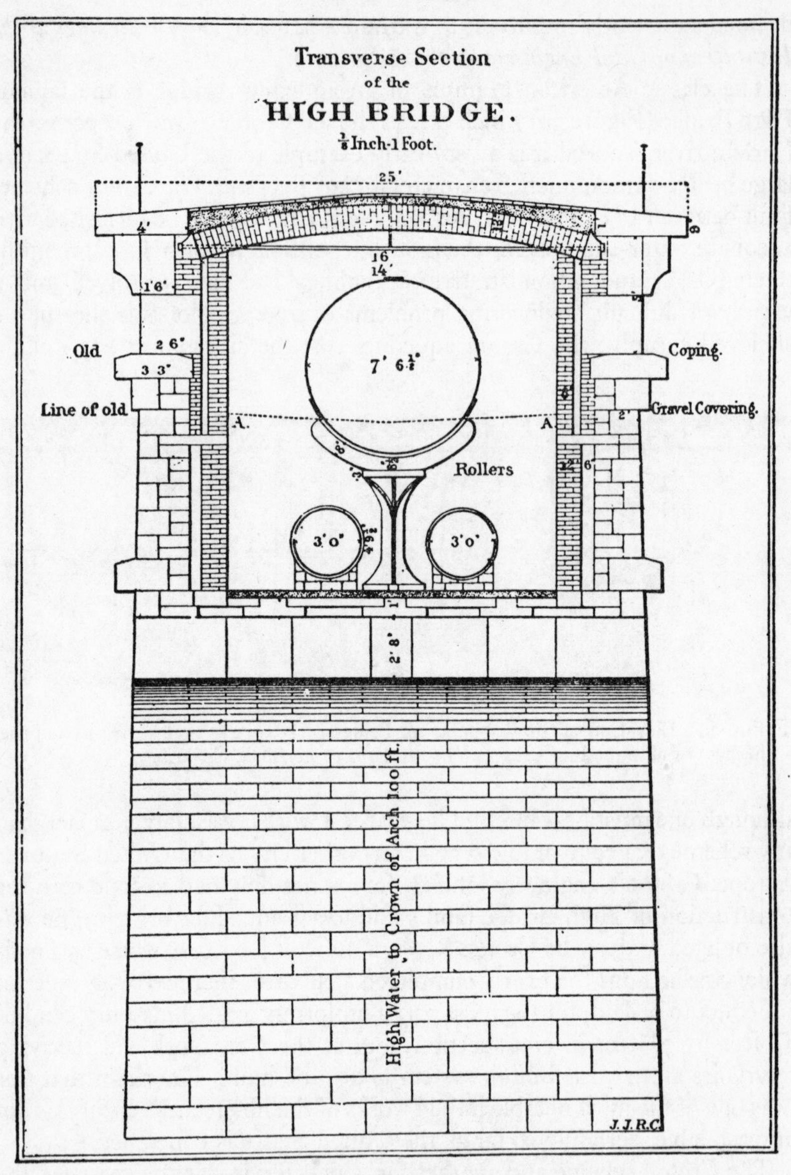

Figure 30 Cross-sectional view of John B. Jervis's High Bridge showing the disposition of the cast iron pipes.

through wooden pipes, and this was augmented by the privately sponsored Aqueduct Corporation at the end of the eighteenth century. Philadelphia, at one time the largest city in the United States, constructed its first water-supply works early in the nineteenth century. Benjamin Latrobe, the engineer-architect responsible for the scheme, pioneered two significant developments: the water of the Schuylkill river was pumped by steam engines and the supply was distributed in cast iron pipes. In 1822 Philadelphia's needs were further met by water-powered pumps installed below the Fairmont dam across the Schuylkill.

In the first half of the nineteenth century no city in the United States was so populous as any of the large cities in Europe. Numbers of people were not the basic reason for introducing water-supplies, and it was the related problems of water quality and disease which exerted the main influence. In the 1790s numerous towns were ravaged by yellow fever, the outbreaks being particularly severe and frequent in New York, Baltimore and especially Philadelphia. Civic authorities were obliged to take decisive action and they resolved, among a variety of measures, to provide more water for public cleansing purposes and better quality water for drinking. Another problem, distinctly more characteristic of American cities than British ones, was the question of fire-fighting. In the eighteenth century many conflagrations had gutted large areas and claimed hundreds of lives in cities of predominantly wooden construction. The issue was to remain prominent in nineteenth-century deliberations over public water, especially in the matter of its distribution within urban areas.

The high charges and inadequate supplies which were characteristic of the early private water undertakings contrasted sharply with the superior service offered by the first publicly run systems. Inevitably, the example which had been set by Philadelphia first of all, and then by the Croton and Cochituate systems in the 1840s, was followed in other places; and in any case the rapidly growing populations of the larger cities, such as New York, Boston, Baltimore, Chicago, New Orleans and San Francisco, were creating demands well beyond the scope of private companies to meet. By 1860 all but four of the sixteen most populous cities in the United States had municipally run water-supply systems.

Invariably the authorities believed that they had planned well into the future; and invariably they were wrong. In the United States just as much as in Europe, the last quarter of the nineteenth century witnessed rapid population growth. Frequently the need for increased supplies was made more desperate by the policy of annexing to large cities the surrounding suburban districts, areas which previously had seen to their own water-supply needs. To compound the problem even further American citizens developed, for various reasons, a much greater thirst than their European cousins; two, three and more times the consumption per head per day was usual.

In the very period when Birmingham, Manchester and Liverpool (and London and Paris for that matter) were constructing new schemes on a scale unanticipated only half a generation before, so New York, Boston, Philadelphia, Baltimore and others also were moving forward an order of magnitude.[3] Every time a new scheme, bigger and better than the last, was inaugurated due pomp and ceremony proclaimed that the water problem was solved. But it never was. New York's Catskill Aqueduct, which took ten years to build, brought water from 120 miles away and as a triumph of engineering over difficult conditions was compared with no less a work than the Panama canal. And yet *before* Catskill was delivering its full supply of 555 million gallons per day, its inability to meet future needs was apparent.

At least in the eastern states there was a good deal of water to be had even when rival cities made plans to utilize the same sources. Imagine how

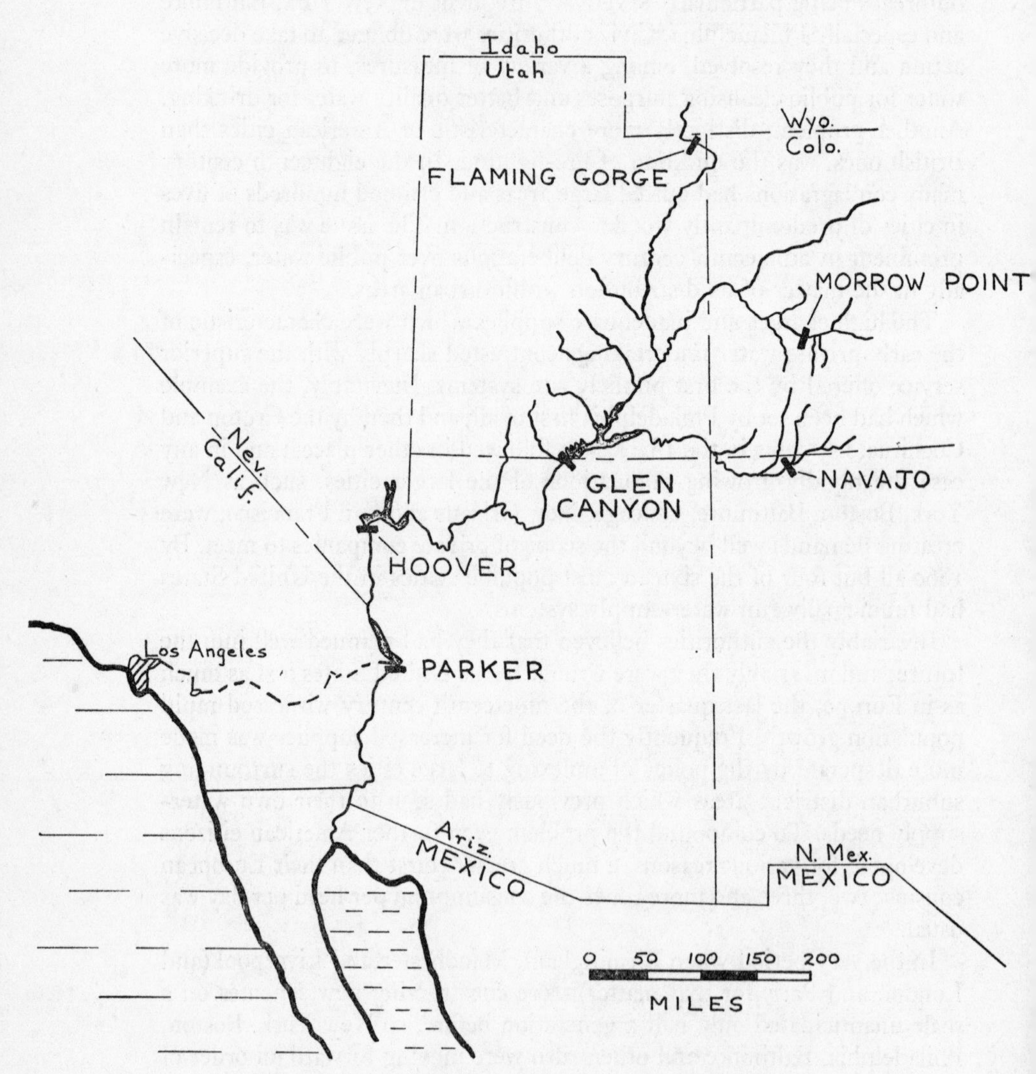

Figure 31 Dams on the Colorado River and its tributaries.

much worse the problem was in the west and on the Pacific coast. In these arid areas not only did different cities compete in the development of water-resources, so did different applications; irrigation was just as big an issue as water-supply and not infrequently either or both were at odds with plans to develop hydro-electric resources. On the other hand, the electrical power made available did benefit the needs of irrigation and water-supply. Across difficult country and over distances measured in hundreds of miles (the Los Angeles aqueduct of 1913 is 233 miles long) power was needed to pump the water against pipe resistance and to lift it over intervening high ground.

The dramatic growth rate of Los Angeles in the early 1920s, a direct consequence of adequate supplies of water for drinking and irrigation, began to pose serious problems at the end of the decade as demand increasingly outstripped supply.[4] Massive engineering was resorted to in an attempt to keep pace. At the beginning of this century the Colorado river (Figure 31), world famous for the grandeur of the scenery through which it passes, was being cursed locally for the variations in its flow and the violence of its floods. Soon after its inception in 1902, the United States Bureau of Reclamation initiated studies of ways and means to tame the Colorado. Many years of laborious topographical surveys, searching hydraulic analysis and detailed structural design led ultimately to the construction, between 1931 and 1936, of the mighty Hoover dam and ancillary works which, by any previous standards, would have been regarded as large in themselves. From the Parker dam, downstream from Hoover, water was piped 242 miles over deserts and mountains to the Metropolitan Water District of Southern California, there to meet the water-supply needs of Los Angeles and its twelve adjacent communities.

The Colorado River Aqueduct supplied 1,000 million gallons of water per day, an unprecedented volume but not in the long run a final solution. Competing power, irrigation and water-supply interests have taken nearly all that the Colorado and its tributaries have to offer. Flaming Gorge, Glen Canyon, Navajo and Morrow Point are the names of a few of the dams which have recently sprung up amid bitter controversy and fierce opposition. Today John Wesley Powell and his contemporaries would hardly recognize the wild river they first explored a century ago. The Colorado is being so completely utilized that no water now flows from the river's mouth into the Gulf of California.

Certain basic elements of water-supply technology – capture in distant catchment areas, conduction along aqueducts and distribution within cities – are as characteristic of recent works as they are of late nineteenth-century projects and indeed of Roman systems. To that extent modern engineering has changed no basic principle; it *has* magnified the scale many times and greatly elaborated the engineering content. In the matter of water-quality, however, the twentieth century has introduced some radically new ideas whose significance could hardly have been foreseen a hundred years ago.

Following the passing of the Metropolis Water Act in 1852, slow sand filtration became obligatory in London and it was rapidly adopted elsewhere. The possibility that the turbid waters of the Mississippi might be cleaned up sufficiently to be drunk by the people of St Louis prompted a study tour of European installations by James Pugh Kirkwood. In his classic *Report on the Filtration of River Waters for the supply of Cities, as Practiced in Europe* published in 1869, one can see for the first time the range of developments under way in Great Britain, France, Germany and Italy. In Germany a component of the improving situation was scientific studies of the theory of filtration. By 1886 Karl Piefke in Berlin had reached an important conclusion. He found that sand by itself was not a cleaner of contaminated water. The effectiveness of the slow sand filter in fact depends upon the formation on its upper surface of a thin layer of algae and plankton capable of a powerful oxidizing, and therefore purifying, action. Above this autotrophic layer a *schmutzdecke*, literally a 'cover of dirt', acts as a fine mechanical sieve; and below there is a deeper region of bacteriological activity which completes the decomposition of organic matter in the filtering water. Combining physical, chemical and bacteriological action the slow sand filter is highly efficient as a purifier but it is also well named: its action is hopelessly slow if large quantities are to be filtered without recourse to massive acreages of filter beds.

The development of methods to filter water rapidly was principally an American achievement. A prominent influence was George W. Fuller, who twice faced the problem of how best to clean the turbid waters of the Ohio River, first of all at Louisville and then at Cincinnati. The essential conclusion of sustained research carried out in the 1890s was that rapid filtration was feasible only if the water was pre-treated. George Fuller realized the crucial importance of adequate sedimentation and a proper degree of coagulation. Sedimentation was an old idea whose value had rather been lost sight of when using slow sand filtration which was so effective by itself. Coagulation as an aid to sedimentation was first used on a small scale early in the nineteenth century, but it only came into vogue in municipal supplies just before 1900. Initially, it was applied as an aid to sedimentation when dealing with the very muddy waters characteristic of many American rivers. Rapid filtration gave coagulation an essential role of its own.[5]

In the nineteenth century a variety of materials were tried as coagulants including metallic iron, iron perchloride and sulphate, lime, and aluminium sulphate. In George Fuller's experiments at Louisville in 1895–6 aluminium sulphate, commonly known as alum, was found to be the most suitable coagulant and many subsequent enquiries confirmed its superiority. It remains a standard coagulating agent in water treatment processing to this day.

By the beginning of the twentieth century rapid filtration was being adopted widely in the United States and was poised to take over in Europe. Notwithstanding its being less efficient than slow sand filtration and despite the need for extensive pre-treatment of the water, the higher filtering speed and smaller size of plant were substantial advantages. Rapid filtration spelt the end for the traditional slow sand methods but at the same time the wide range of cleansing actions which were a natural feature of slow sand filters somehow had to be reproduced in the rapid filter. Coagulation was only one of a variety of physical and chemical processes developed to rectify specific features of contaminated water. Water *treatment* has been a major theme of twentieth-century developments.[6]

Disinfection, or germ-killing, has employed many agents including copper, silver, ozone, ultra-violet light and heat. All are effective to varying degrees but none are so potent or so universally used as chlorine. Tentative and ill-designed nineteenth-century experiments led eventually to working-scale chlorination plants being installed in the United States, Belgium and England around 1900. None, however, were as effective in establishing the value of the process as the plant built by the Jersey City Water Works at their Boonton Reservoir in 1908. Chlorination was then taken up rapidly in the United States but with less conviction elsewhere. Germany was particularly slow, only 30% of city water-works using the process as late as 1941. Chlorination was only spasmodically practised in Great Britain until a serious outbreak of typhoid at Croydon in 1937 caused a marked shift of opinion in its favour. The cost of chlorination was accepted as a necessary insurance of people's health, and the objection that such sterilization imparted a disagreeable taste to drinking water was overruled.

Nowadays, chlorination, often in more specialized forms than the original process using simple injections of chlorine gas or sodium hypochlorite, is universal while the more expensive but very powerful ozonization process is practised as well.

As distinct from the contamination of water by bacteria and viruses, pollution by microscopic plant and animal life (algae) is a problem whose solution was first set out in 1905. The presence of algae is not altogether a bad thing since they assimilate carbon from organic compounds and thereby liberate oxygen – and they perform a vital function in slow filtration techniques. But in abundance they will taint water when they decompose, and clog filters with their remains. In 1905 George T. Moore and Karl F. Kellerman of the U.S. Department of Agriculture showed that copper sulphate, in concentrations no greater than one or two parts per million, was the answer.

The presence of the carbonates and bicarbonates of both calcium and magnesium cause water to be 'hard'. Although hard water is excellent for drinking it is difficult to make it lather in the presence of soap and much more of a drawback is its ability to 'scale' the insides of kettles, boilers and pipelines at high temperatures. The problems of hard water were already crucial in the nineteenth century and users of steam power – ship-owners, factories, railway companies and electricity generating stations – all had a substantial interest in efforts to soften water. Before 1900 only a few municipal undertakings took steps to provide water-softening; the plant built in 1888 at Southampton was for some years the biggest in the world but even so it only treated two million gallons per day. Nothing substantial was realized until the twentieth century and it remains obscure as to why the basically simple processes required were not taken advantage of sooner. Presumably matters affecting health took precedence over those which were a question mainly of convenience. Two softening processes are now established. The first adds lime and soda to the water in order to convert soluble hardness compounds into insoluble precipitates which can be removed by coagulation and filtration. The more modern zeolite process is of German origin and established itself in the 1930s. Based on electrolytic action, the zeolite process produces no sludge and this is an advantage.

In order to maintain in drinking water a suitably high level of dissolved oxygen (and a proportionately low level of carbon dioxide) aeration has been practised for nearly two centuries but only in the last fifty years or so have the precise requirements of the process become well understood. Two basic techniques are now used. Cascading water down a series of steps is cheap and easy but only moderately efficient; spraying water into the atmosphere achieves a much better result but only if the cost of pumping through jets can be borne.

Even a brief outline is bound to illustrate how varied and numerous are the physical and chemical techniques which in modern times have become part and parcel of water treatment processes. The physical, chemical and bacteriological purity of drinking water can now be assured, although the purification process must not be taken too far. Even though it could be done, at a price, the aim is not to produce an unadulterated compound of hydrogen and oxygen; that would be a tasteless and very objectionable liquid quite unacceptable for drinking. The aim is to supply water which is medically safe but also aesthetically acceptable in terms of taste, colour and odour. It is not an easy result to achieve and the required technology is complex. It is an interesting thought too, that while the capture, conduction and distribution of water involves civil engineering on a massive scale, water treatment is contrastingly delicate and carried out with laboratory precision. Moreover, no two sources of water are the same in terms of the treatment they require. All must be analysed individually and treated accordingly. Vigilance is essential, particularly in an era when the natural impurities in water sources are, if anything, less threatening than man-made pollution which varies continuously, and occasionally as a result of an accident or some gross irresponsibility, can quite suddenly produce a potentially lethal situation. Particularly troublesome contaminants of wholly recent origin are synthetic detergents, certain effluents from the plastics industry and radioactivity.

The problem of water purification is essentially a question of removing all manner of objectionable, harmful and undesirable ingredients; in origin these substances can be natural, man-made or the wholly artificial result of some deliberately induced purifying process. There is, however, another side to water treatment. It involves the concept that water should have substances added rather than removed – that, in short, water-supplies should be used as a means of communal medication. The issue is, and always has been, highly controversial.

The fact that goitre is attributable to an iodine deficiency prompted a few attempts at iodization of water-supplies in the 1920s and '30s. None of these efforts (a few in the United States, and one in Derbyshire) lasted long. Various factors conspired to halt progress almost as soon as iodization was under way. Public apathy was influential and so was determined opposition by professional and other groups who in general were opposed to the principle of medicating water-supplies in any way or under any circumstances (not infrequently chlorination was similarly resisted). The favoured alternatives were iodized sweets given to school-children and the use of table and cooking salt high in iodine content. These methods, it was felt, offered people a degree of choice much preferable to the sinister notion of forced medication by local authorities.

Dental caries can be prevented by adding the correct quantity of fluorine to water-supplies. The required concentration is very low (1–2 parts per million) and does not have to be exceeded by much to produce dental fluorosis, a disease which gives teeth a mottled and displeasing appearance. The idea of mass fluoridation continues to be controversial and the opposing factions cite various arguments for and against. Dental caries is by far the most serious dental problem the world over and costs millions of pounds each year to treat. Fluoridation, by contrast, is cheap and effective and easy to arrange. On the other hand, there are no options. Young and old alike are dosed and there is evidence that only the young benefit. No matter how weak the initial concentration of fluorine may be it is suggested that dedicated drinkers of water might ultimately accumulate so much fluorine in their bodies that dental fluorosis would be bound to develop. Opposition to the very idea of mass medication is powerfully expressed by those who are fearful that here is just one more instance of the erosion of the individual's right to choose. Nevertheless, fluoridation is already well established in many countries, especially the United States. How far the practice will be allowed to develop and whether it will eventually lead to yet more centralized control of people's health, or worse, remains to be seen.

Nowadays water-supply is a highly developed and elaborate technology, a striking contrast to the situation which prevailed even one hundred years ago. Water treatment techniques have been refined to such a point that there is practically no water, no matter how polluted, that cannot be made potable at a price. What that price is, however, is another matter and the economics of water treatment is fast becoming a very serious issue. The civil and mechanical engineering components of water-supply are well worked out and thoroughly understood. The biggest water-supply schemes in the United States offer convincing proof that so long as someone will foot the bill for the structures, pipelines, treatment plants and machinery that must be set up, enormous quantities of water *can* be supplied provided that sufficient and suitable sources of water continue to be available. But will they, and what are the rival claims of irrigation and power? The world's water problem will be examined in Chapter 16, 'Man and Water'.

Part III

Water into Power

II

The Water-Wheel

IN HIS VALUABLE and wide-ranging studies of ancient technology, R. J. Forbes argues[1] that prime movers are the 'keystone of technology', that above all else it is the availability of energy which controls and determines the development of technology generally. On this basis, Forbes proceeds to the idea that the history of technology is characterized by five 'power stages'.

In the first, the simple nature of the energy source greatly belies its significance; over a period of several thousand years man achieved great advances and wrought important changes to his way of life and his environment using only his own muscles. There then followed a phase in which the muscles of animals were added to those of men, with the result that there was a marked increase in the overall amount of energy available. The use of wind- and water-power marks the third stage, of great significance since it not only made available much greater quantities of power but greatly increased its concentration in any single power-producing unit. Beginning with the reciprocating steam engine and gradually evolving so as to embrace the internal combustion engine and the steam turbine, what might well be termed the heat engine stage corresponds broadly to the Industrial Revolution. The opening of the fifth phase, the nuclear phase, is a feature of modern times.

These five divisions do not, of course, represent complete replacement of one mode of power by its successor. The muscle power of men and animals, for instance, continues to be globally exploited but not for centuries has it made more than a marginal contribution to the overall picture. Nor has nuclear power yet made inroads into the predominating roles of coal, gas and oil. Nevertheless, Forbes' sequential classification is useful in depicting five phases of innovation in prime movers, each one of which has tapped new power sources leading to a corresponding increase in the amount of power actually available to a few, but potentially available to all.

Within this five-stage classification water-power occupies an interesting position. It was the first form of inanimate power ever to be exploited, some 2,000 years ago, and yet unlike some of its later rivals, such as the reciprocating steam engine, the windmill and the gas engine, all of which had their heyday and were then superseded, hydraulic motors have remained an important power source right down to modern times. Nor is water an energy resource which is susceptible to depletion, something which cannot be said of fossil fuels (coal, gas and oil) whose steady diminution poses, for the future, a serious problem to which water and nuclear power may well provide the only solution.

The origins of water-power and the nature of the earliest water-wheels are extremely obscure. However, a handful of references from the first two centuries preceding the Christian era are indicative of some degree of useful application and at least regional recognition of the idea of water-power.[2] A second-century papyrus and Lucretius' *De Rerum Natura* (mid first-century) both suggest the use of the noria while the Greek geographer Strabo (*c.* 60 B.C.–*c.* 25 B.C.) and a certain Antipater, of doubtful provenance, mention the application of water-power to corn grinding. Lack of information as to the form of the wheels has not prevented the widespread acceptance among modern writers that Antipater's wheel drove a vertical axle (Figure 32). Herein lies the origin of the term 'Greek mill' so frequently used to denote a gearless corn mill driven by a horizontal wheel; but in fact the appellation is quite impossible to justify.

The writings of Vitruvius (*c.* 25 B.C.) are the earliest reliable source of information on water-power. Chapter V of Book X of the *Ten Books on Architecture* consists solely of the following:

> 1. Wheels on the principles that have been described above are also constructed in rivers. Round their faces float-boards are fixed, which, on being struck by the current of the river, make the wheel turn as they move, and thus, by raising the water in the boxes and bringing it to the top, they accomplish the necessary work through being turned by the mere impulse of the river, without any treading on the part of workmen.
>
> 2. Water mills are turned on the same principle. Everything is the same in them, except that a drum with teeth is fixed into one end of the axle. It is set vertically on its edge, and turns in the same plane with the wheel. Next to this larger drum there is a smaller one, also with teeth, but set horizontally, and this is attached (to the mill-stone). Thus the teeth of the drum which is fixed to the axle make the teeth of the horizontal drum move, and cause the mill to turn. A hopper, hanging over this contrivance, supplies the mill with corn, and meal is produced by the same revolution.

Vitruvius's descriptions of these two applications of the river-driven

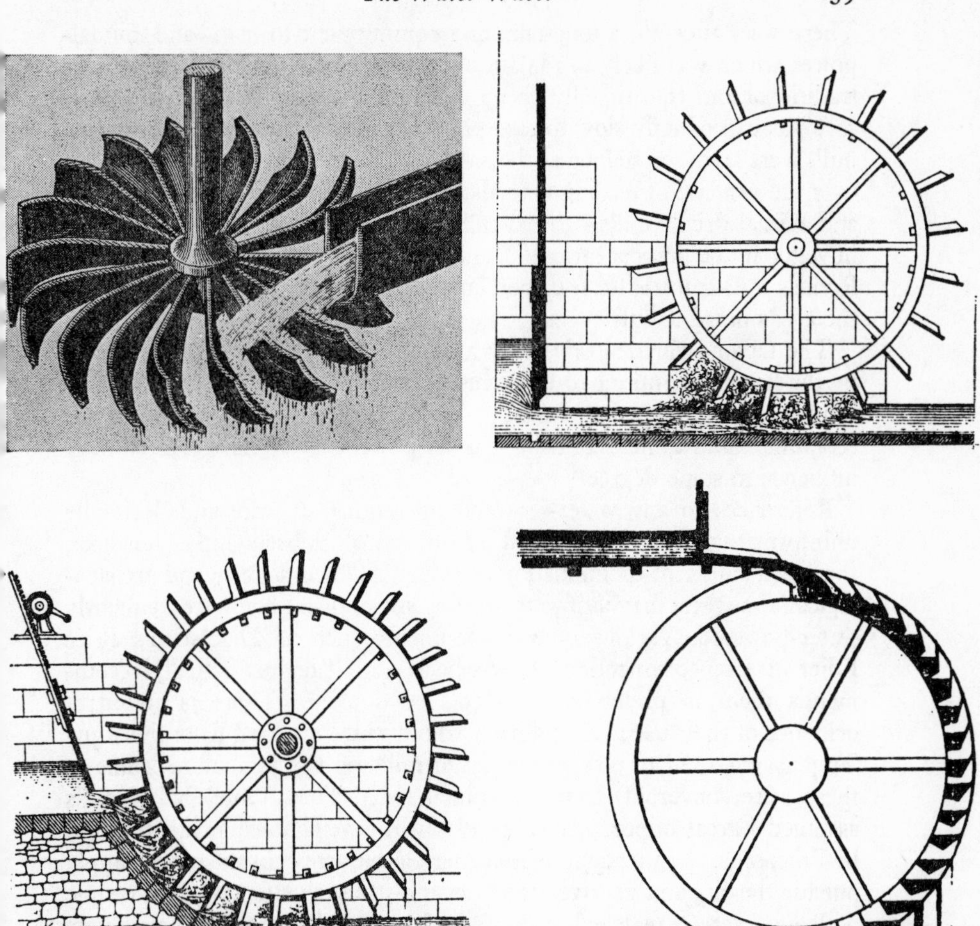

Figure 32 The various water-wheel configurations: *top left*, the horizontal water-wheel; *top right*, the undershot water-wheel; *bottom left*, the breast-wheel which is referred to as low-, medium-, or high-breast depending on the level at which it is supplied; and *bottom right*, the overshot water-wheel.

undershot water-wheel (Figure 32) are lucid and unambiguous. He makes no claim of originality or novelty for the undershot water-wheel and its regular, but not necessarily widespread, use in the last century B.C. is implied. On the other hand he is apparently not acquainted with the horizontal wheel, or Greek mill. In view of his apparent familiarity with Eastern Mediterranean technology, the latter omission is notable but how significant it is not easy to assess.

The water-wheel was slow to be adopted in Roman times. This is not necessarily indicative of any inherent defect in the principle of the machine but rather it is a comment on prevailing social and economic conditions.

There was, after all, a long-standing commitment to man- and animal-power which was likely to remain effective in societies where these were traditional and traditionally cheap to employ, and where change of any sort was exceedingly slow to take place anyway. Presumably man-driven mills were favoured so long as the cost of keeping slaves offered advantages over the expense of feeding animals. But this situation evidently changed and animal-driven mills – the so-called Pompeian mills – were very much in vogue in the first centuries B.C. and A.D. When Caligula commandeered Rome's mill-animals for transport in A.D. 39, he immediately jeopardized the city's bread supply.

The factors which slowly brought about the adoption of water-power by the Romans continue to be discussed. The rise of Christianity with its implicit rejection of slave labour, a developing labour shortage and the economic disadvantages of using animal-power may all have exerted their influence to some degree.[3]

References to any water-powered operations are rare and decidedly uninformative before the third century A.D. Subsequently, however, there is a sufficient accumulation of evidence, documentary and archaeological, to suggest that both water-raising and corn-milling were frequently water-powered. Various official documents such as Diocletian's Price Edict of A.D. 301 mention water-wheels and Palladius actually recommends them in preference to slaves or donkeys. There is abundant evidence of their use in Rome itself where water diverted from the Aqua Trajana was used to power a group of mills on the Janiculum; some of these were converted slave and animal mills. That Rome's water-mills assumed a great importance is borne out by a well known incident related by Procopius.[4] In A.D. 537 the besieging Gothic army cut the water-supply outside the city and deprived the grist-mills to such effect that the people of Rome were threatened with starvation. A solution was immediately engineered by Belisarius, who set up on the Tiber a series of undershot water-wheels mounted on barges, the first known reference to floating mills. It is of some significance that by the sixth century the Romans' commitment to water-power was so complete that the failure of one group of water-mills led to the construction of an alternative set rather than a recourse to slave or donkey mills.

In Rome itself virtually no remains of such early water-mills have ever been found. In other places, however, traces have been unearthed from time to time. In 1938, for instance, remnants of a 2-metre undershot wheel for corn-grinding exactly in the Vitruvian manner were discovered at Venafro near Naples. Much more impressive is the early fourth-century flour factory located at Barbegal near Arles in southern France, an installation comprising no fewer than sixteen water-wheels capable of producing some 28 tonnes of flour per day when running at their maximum capacity. Such a vast potential – if not actual – output must lead one to suppose that the Barbegal mill was a central corn-grinding plant meeting the needs of a large area such as Narbonensis. Or conceivably it was for military supplies in the manner of a similarly large water-powered mill at Tournus, north of Lyon, which sustained the armies of central and northern Gaul.[5]

Only two illustrations of water-mills have survived from antiquity. A mural in the Roman catacombs depicting an overshot wheel dates from

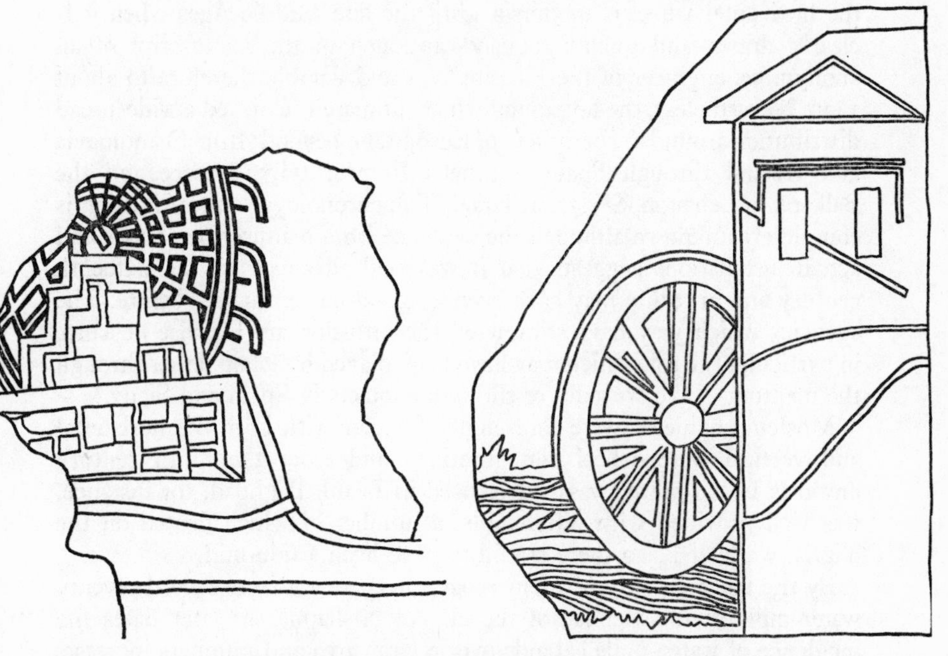

Figure 33 Two early mosaics depicting water-power: *on the left* a noria of the second century A.D. at Apomoea in Syria; and *on the right* a fifth-century Byzantine undershot water-wheel.

the third century and a fifth-century Byzantine mosaic shows an undershot wheel (Figure 33). The remains of a fifth-century overshot wheel were discovered in the Athenian Agora in 1936. The earliest use of water-power in Britain dates from Roman times. Three third-century examples have been identified on Hadrian's Wall and remnants have been located in Lincolnshire, Essex and Gloucestershire. Evidently, towards the end of the Roman era the distribution of water-wheels was widespread but how numerous they were is impossible to say.

The available evidence suggests that the horizontal water-wheel was unknown to the Romans. In fact, the machine's early appearance in Denmark and subsequent application in Norway, Sweden, Scotland and Ireland is suggestive, on balance, of a far northern European origin rather than a Mediterranean one. The beginnings of the use of horizontal water-wheels remain highly mysterious, and certainly more so than for the vertical wheel for which at least we have Vitruvius's word.[6]

Bearing in mind the importance of the horizontal wheel in Ireland at later and verifiable dates, it seems likely that this was the type used there as early as the seventh century A.D.; and confirmation is to be found in the description of a water-mill in a collection of tracts called *Senchus Mor*, first written down in the eighth century. Otherwise clear-cut evidence of the horizontal wheel is unknown until the late Middle Ages when it is clearly drawn and unambiguously captioned in the manuscript of an anonymous engineer of the Hussite War and datable therefore to about 1430. Nevertheless, the horizontal wheel ultimately achieved a widespread distribution around the periphery of Europe and beyond; from Scandinavia and Ireland through Spain, southern France, Italy, Greece and the Balkans to Lebanon, Syria and Israel. The chronology of this diffusion is completely unknown although the evidence from manuscripts, books and actual installations suggests that it was well advanced by the sixteenth century and in reality may have been long – even centuries – before. The agencies which may have stimulated the diffusion are likewise obscure, in particular the role which may have been played by Islam either through the medium of the crusades or through contacts in Spain and Sicily.[7]

Moslem engineers were thoroughly familiar with both the horizontal and vertical water-wheel configurations and from the ninth century onwards hydro-power was widely used in Islam. Baghdad, for instance, was well provided with water-mills, a number of which floated on the Tigris, while at Basra there is said to have been a tide-mill.

By the tenth century Moslem geographers were able to record seventy water-mills in the vicinity of the city of Nishapur. At later dates the incidence of water-mills extends over a large area and numbers increase. At the beginning of the thirteenth century al-Marrakushi mentions 300 mills in Marrakesh alone. As in Europe, water-power in the Islamic world was applied at first to water-raising and corn-grinding but was gradually extended to such work as sugar-cane crushing, fulling and paper-making. An interesting variation was the occasional use of dual purpose water-wheels: norias used to drive the grinding stones of grist-mills.

In an effort to improve the performance of horizontal and undershot water-wheels Islamic engineers were among the first to use river dams, generally not very large, to increase the velocity of flow and provide a measure of flow control. Hydro-power dams were especially numerous on rivers such as the Karun, Kur, Helmund and Oxus in the eastern caliphate and on the Guadalquivir in Spain.

The Guadalquivir was a river of some importance in Moslem Spain; a large proportion of the population of al-Andalus lived on or near its banks and it connected two important cities, Cordoba and Seville, with the sea.

Apparently a good deal of power was derived from the river by means of water-wheels mounted on the banks, on floating barges and within or at least supplied from dams. One of these dams still stands across the river at Cordoba just below the Puente Romano. Until quite recently its thre-mill houses – al-Idrisi says that each contained four water-wheels – still functioned, but much changed from their original form.

The early utilization of water-power in Spain, then, is well established in extent if not in detail. How much influence Spanish Moslem practices could have exerted on the rest of Europe is uncertain. New ideas may conceivably have been introduced, water-wheel styles may have been diffused and the spread of water-powered industries such as paper-making may have been encouraged. But for all that, there is substantial and quite independent evidence of relatively widespread hydro-power developments beyond Iberia some centuries before the Reconquista.

In the Middle Ages the water-wheel experienced a certain amount of basic development. Among vertical water-wheels, the undershot variety clearly predominated and it is not until the fourteenth and fifteenth centuries that we find crude illustrations depicting the use of the overshot wheel.[8] In its simplest form with the blades simply dipping into a river, the undershot wheel is a clumsy and inefficient machine. Its speed of rotation is limited to some proportion of the river's own stream velocity, usually not very high, while a rise or fall of water level will tend either to drown the machine or leave it high and dry. Medieval engineers took a few steps to offset these inherent defects. Floating mills were one solution which achieved a certain degree of recognition since this is certainly an arrangement which will always overcome the problem of changes in river level.

The oldest illustration of a floating mill, in a French manuscript of 1317, shows the boats tied up between the piers of a bridge across the Seine in Paris, an arrangement which was in use as early as the twelfth century (actually in the reign of Louis VII, 1137–80). This technique was clearly an attempt to improve the mills' performance by utilizing the increased velocity of flow as the river streamed through the bridge. From this the logical idea of providing a fixed water-wheel in the openings of a bridge was but a small step. Sure enough one finds the first evidence of this in Paris early in the thirteenth century.[9] Subsequently the undershot wheel located under a bridge was to enjoy a certain vogue and was a prominent feature of several seventeenth- and eighteenth-century water-pumping installations in large cities (Chapter 8).

The floating mill with its advantages of mobility and adaptability to changing water levels achieved a widespread distribution. It remains uncertain, though, how numerous they were in comparison with fixed wheels. The conventional river-driven undershot wheel continued to be widely used and medieval engineers took steps to improve its performance and to provide means of control. River dams were a basic element in these developments. Long overflow weirs built across rivers provided a sensibly constant upstream water level in all but extreme flood conditions. One or more sluices located between the ponded water and the wheel provided a degree of control of the velocity and quantity of flow, while to some extent the conditions of impact between water and wheel were improved. The use of dams meant that for the first time it became commonplace to drive vertical wheels from an artificially contrived head of water, something which in principle had always been a feature of the horizontal water-wheel.

Outside Spain, the damming of the River Leck around A.D. 1000 to supply water-mills at Augsburg is the earliest clear reference in Europe to this new development. In 1171 the Count of Toulouse gave the Bishop of Cavaillon the right to build a hydro-power dam on the River Durance, the River Drac at Grenoble was dammed in 1191, and the Garonne at Toulouse in the thirteenth century. In his account of the career of Philip Augustus, William the Breton says that the king, when besieging the town of Gournay near Beauvais, hastened its surrender by breaking the dam which fed the town's mills.

Once medieval engineers had arrived at the concept of artificially contrived falls of water the opportunity of using the overshot water-wheel presented itself more forcibly. Since an overshot wheel is supplied at the top and discharges at the base, the overshot wheel can, in principle, be as large in diameter as the height of the dam used to create the fall of water. A manuscript illustration establishes the use of the overshot wheel at least as early as *c.* 1338. Such subsequent information as we have, however, seems to indicate that the overshot wheel was not used so widely as other types for some centuries.

The considerable evidence of the extent to which water-power was being used by the eleventh century is an indication that in fact, in the preceding centuries, much had been achieved, even though documentation is sparse. In 1086 Domesday Book listed 5,624 mills in England alone and this large number must surely be regarded as the accumulation of a long period, certainly from as far back as 762 when a Saxon charter of Ethelbert makes reference to a monastic mill near Dover, the earliest English example. Nor is there any reason to believe that England was particularly advanced in power technology. What Domesday Book tells us about England in particular can probably be taken as typical of quite large parts of Europe in general.

With the twelfth and thirteenth centuries the commitment to a labour-saving power technology increased and for a few places its extent is graphically attested by statistics relating to the rate at which water-power installations were constructed.[10]

Of perhaps greater interest is the vast array of technical operations to which water-power was applied for the first time. The list in fact is too long to enumerate in any short account but one can summarize by remarking that fulling, iron working, saw-milling, paper making, tanning, wire drawing, boring wood and metal, grinding various materials, olive pressing, bark and ore crushing and some finishing operations to tools and metal were all widely mechanized in the later Middle Ages and in most cases the earliest instances, even the first illustrations in a few cases, are known before 1300.[11] Of particular significance is the extent to which water-power formed the basis of the early development of mining and metallurgy in Central Europe.

Throughout this period of such far-reaching change in the field of mechanical power it is likely that the horizontal water-wheel remained confined to corn milling; its relatively meagre performance would have prevented its use for heavy work while its gearless simplicity rendered it particularly suitable for this one operation, especially if demand was small and intermittent. The fact that Domesday Book and other documents and charters feature a distinct category of corn mills whose yearly rental is of a lower order, may perhaps indicate the number of horizontal mills in use. But it was the vertical wheel, undershot in the main but sometimes overshot, which was most suitable for the various heavy applications introduced between 1000 and 1800. This type of wheel was much more amenable to increases in power (by raising its diameter or width or both) and geared drive was a long-standing feature of the type. Even then, however, there were a variety of problems to tax the ingenuity of medieval mechanical engineers. While the continuous rotary motion of a water-wheel, geared up or down, is directly applicable to some operations – e.g. boring, certain grinding jobs and milling – for the very important applications to fulling, iron working (forge hammers and furnace bellows), sawing, or crushing, rotary motion must be converted to reciprocating motion. The tappet was the device most frequently used to achieve this. Projections on a rotating shaft (not infrequently the water-wheel's own axle) were made to strike the tips of pivoted levers, or pegs let into a vertical rod. Very often some sort of spring loaded return mechanism had to be provided since the tappet alone is not a positive drive. Surprisingly, the use of that most commonplace of modern mechanisms, the crank and connecting rod, was virtually unknown until the end of the fifteenth century but thereafter examples of its use in water-powered machines are legion.

Along with certain agricultural innovations, the power revolution of the later Middle Ages is the most notable aspect of the whole of medieval technology. It can be observed too that the commitment to new sources of power was so complete that people unable to harness rivers and streams were not to be thwarted. References to tidal power appear first in the eleventh century and occasionally in the twelfth. Subsequently tide-mills have a more or less continuous history to modern times and although they cannot be regarded as a major influence they are indicative of an attitude. In places where there was no power to be had from fresh or sea water, medieval technicians looked to another natural force, the wind. The windmill, whose origins are about as obscure and tantalizing as those of the water-wheel, was an important innovation of the Middle Ages, a highly significant and ingenious manifestation of the success of the power revolution.

The successful attempt to replace the labour of human beings by means of machines, at the same time producing much increased quantities and concentrations of power, was bound to have far-reaching effects. The very location of industry was controlled in no small way by the availability of water, a factor which some authorities have claimed acted to the advantage of northern European countries generally. The centres of certain industries were obliged to shift. In England there was a general movement of the woollen industry from the south and east to the north and west where greater reserves of water-power were available for fulling mills. The feasibility of larger corn-mills led in a diffuse way to a concentration of flour grinding facilities from which emerged the figure of the professional miller and that famous medieval institution, the manorial mill Here alone, in theory at least, society was supposed to pay to have its flour ground to the sole benefit of the lord of the manor, for whom the miller acted as agent and as a result came to be generally mistrusted and disliked; Chaucer's portrayal is unmistakable evidence of this.

Among the new operations and processes dependent on water, those concerned with mining and metallurgy were probably the most significant, certainly in the long run. From the outset water-power was of critical importance in launching German and Hungarian metal mining exploits, which eventually were to exert a marked influence on European technology generally. One metallurgical result of hydro-power stands out from the others; the use of water-wheels to blow iron smelting furnaces. A new generation of blast furnaces developed such high temperatures that iron could be melted. As a result rates of production were generally increased and cast products, notably cannon, became possible for the first time.

For the fifteenth century the means to appraise late medieval and early Renaissance water-power technology is greatly aided by the existence of a number of illustrated manuscripts by German and Italian engineers.[12] Notable among these are the compilations of Konrad Kyeser (*c.* 1405), an anonymous German engineer of the Hussite Wars (*c.* 1430), Mariano di Jacopo Taccola (*c.* 1450) and lastly the voluminous notes of Leonardo da Vinci. Depictions of undershot, overshot and horizontal water-wheels are numerous. Many of the technical operations already discussed are featured and at last the horizontal wheel is shown coupled to geared drives. However, it will be expedient at this point to put the fifteenth-century horizontal wheel to one side and leave until later its subsequent development along with the difficult question of the origins of the water-turbine.

Hydraulic studies figured prominently in the work of Leonardo da Vinci and he has a good deal to say, much of it original, on water-power.[13] His are the first written thoughts on what might be called hydro-power theory. He dispenses brusquely and categorically with the long held view (recurrent in the work of Leonardo's contempories) that water-wheels offered a solution to the quest for perpetual motion. 'Descending water will never raise from its resting place an amount of water equal to its weight', he says. At the same time he appreciated that water's potential to do work depended on its fall, with the proviso that much of the energy theoretically available would be lost as frictional resistance in the wheel and the machinery it drove. Undershot wheels he regards as most efficient when the stream of water strikes the blades at right angles. More interesting is his observation that the overshot wheel is much the most efficient arrangement, a conclusion which only became widely known two-and-a-half centuries later as a result of John Smeaton's classic experiments. With considerable perception Leonardo saw too that the overshot wheel works by weight alone and depends not at all on any impact between the moving water and the buckets of the wheel. For all his theoretical insights, however, Leonardo was restricted to statements of principle; nowhere does he attempt to quantify head, pressure, power, velocity or any of the other parameters whose relevance he sensed.

Leonardo's manuscripts are full of practical propositions concerning not only water-wheels but a vast range of applications as well. Here for instance we find the earliest known reference to the breast-wheel, an improved form of the undershot wheel which was ultimately widely used (see Figure 32, page 139). Conscious that submersion in its tail race would detract from a water-wheel's performance, Leonardo suggests various remedies including going so far as to provide a device with which the wheel could be raised and lowered bodily. Various ideas are proffered in an effort to retain water in the buckets of the overshot wheel until the last possible moment before discharge.

Although Leonardo's writings teem with ideas and propositions, we are uncertain how much was due to Leonardo's own inventiveness and how much is a record of his own and other people's observations (the manuscripts recently discovered in Madrid, for instance, include a lengthy bibliography suggesting that Leonardo was in fact familiar with many contemporary writings). Nor can the extent of Leonardo's influence be judged at all accurately. While it is true that many of his ideas eventually turn up in practical engineering, it by no means follows that familiarity with Leonardo's writings is the reason for this.

From the sixteenth century onwards the three forms of vertical water-wheel – undershot, breast and overshot – experienced no basic developments for nearly 300 years. Steadily, and in order to meet increasing power requirements, it is true that wheels did tend to become larger and a sustained commitment to water-power certainly caused numbers to increase; but that is all. On the other hand one can detect a tremendous variety of applications of water-power and occasionally the appearance of a new one. In some industries one gets the impression that its use was universal. Engineers now turn to the water-wheel as a matter of course and to find its use recommended in the sixteenth century for such trifling work as ribbon-weaving, shaping vases and polishing gemstones is evidence of this.

Among various sixteenth-century printed books on technology, water-power and its applications comprise the largest single topic. These books range from practical manuals such as Biringuccio's *Pirotechnia* (1540) and Agricola's *De Re Metallica* (1556) to the fanciful but intriguing 'Machine Books' of Agostino Ramelli (1588) and Jacques Besson (1579). In *De Re Metallica* can be found the whole range of water-power applications in mining and metallurgy, at least so far as the more advanced state of this industry in Germany and Central Europe was concerned. Overshot wheels (including a reversible type) are just as usual as the undershot variety and both are shown pumping water, raising spoil, ventilating the mines, crushing and grinding the ores and blowing the furnaces. For each of these operations two and sometimes three variations are described. Moreover Agricola only rarely sees fit to remark on the novelty or originality of his machines, suggesting that the bulk of *De Re Metallica* is an account of widely known and routine applications. By comparison the writings and drawings of Ramelli, Besson and such seventeenth-century disciples as Zonca, Strada and Branca are excessively novel and not infrequently absurd. Nevertheless, the emphasis on hydraulic matters in general and water-power in particular is probably a sure guide to a widespread interest in and acceptance of such matters.

By the seventeenth century the growth of a number of manufactures had produced centres of such size and influence that the total life and prosperity of a region centred on its complex of water-powered machines. Among such areas was the Oberharz, the great German mining and metal working area in the Harz mountains south of Brunswick. Evidently it was here, a century before, that Agricola gathered some of his material as he journeyed through central Europe. In Russia there are several examples, notably the famous metal working centre of Tula which was established towards the end of the seventeenth century and greatly expanded at the beginning of the eighteenth century at the instigation of Peter the Great in order to meet the country's armament requirements. By 1800 the Tula plant was one of the biggest gun manufactories in the world. Metallurgical processing also prompted some extensive water-power developments in the New World; one such example was particularly impressive. High up in the Andes near the town of Potosí in Bolivia, the conquistadors set up extensive hydraulic installations featuring 132 water-wheels to crush and process silver ores.

From the Middle Ages onwards, whenever the demand for water-power could not be met by normal run-of-river supply, with or without diversion dams, water-storage was undertaken. In most cases such provision was on a small scale and the mill-pond became a characteristic accessory of water-powered plant throughout Europe. It was in the nature of the evolution of hydro-power, however, that in time storage on a more ambitious scale would be attempted. The first European countries to take this step were Spain and Germany, and for the first time the construction of big dams became a feature of hydro-power engineering.[14]

The earliest Spanish power dams can be traced back to the sixteenth century; none, however, are as notable as the eighteenth-century dam of Albuera de Feria. Built in 1747 to power a grist-mill near Badajoz, it is one of the first dams to contain a water-wheel within the body of the structure and, moreover, it is the oldest example of a dam in which fully developed buttressing is an integral part of the structure. In the second half of the eighteenth century Spanish-built buttress dams were the first to impound water for power in Mexico.

Power dams in Spain were an easy progression from decades of experience in irrigation. In Germany, however, it was the need to sustain an all-the-year-round supply of water for mining and metallurgical processes which in the sixteenth century prompted the first big dams for any purpose. The Oderteich dam in the Harz was a culminating point of the early developments. Standing to a height of 22 metres it is the first recorded instance of a rock-fill dam in the world.

Water-storage dams figured prominently in eighteenth-century power plants in Russia, notably in the Urals and the Altai region of Siberia.[15] An important pioneer of their construction was Koz'ma Frolov (1726–1800), a great figure in the annals of eighteenth-century Russian engineering. The most famous monument to his skill as a dam builder is the 17-metre-high Smeinogorskii dam in the Altai. It was built to high standards and on eminently sound principles in the 1780s. Notwithstanding the savage conditions prevailing in central Siberia and the lack of anything but very primitive constructional equipment, Frolov and his workers placed no less than 100,000 cubic metres of earth into an embankment which has survived to the present time. The reservoir formed was the largest artificial lake impounded solely for water-power requirements until well into the nineteenth century. The power was used to drive mine machinery and, more important, to pump the deep level workings of the mine.

12

The Industrial Revolution

ALTHOUGH MANY OF the examples of eighteenth-century water-power discussed above were applied to industrial processes, they did not contribute to any rapidly accelerating process of industrialization. That privilege was reserved for engineers in Great Britain. Just as drainage was a major problem in the mines of Germany and Russia, so it was in England. It was this issue which stimulated in Great Britain experiments in the use of steam power and led Thomas Newcomen to develop the atmospheric pumping engine, which in turn, by way of James Watt's introduction of the separate condenser, produced the most celebrated machine of the Industrial Revolution – the reciprocating steam engine. So admirably suited was steam power to mine pumping – it was after all its *raison d'être* – that water-power was little used in this field in Great Britain; but it was tried. In some industries, lead mining in the Pennines for instance, water-wheels were able to compete in the face of the high cost of erecting steam engines and transporting coal, while as late as 1854 in the Isle of Man the impressive 'Lady Isabella', the largest if not the most powerful water-wheel ever constructed in the British Isles, was also a mine pumper.

For a period, roughly 1750 to 1825, some miners employed water-pressure engines.[1] This curiously overlooked breed of prime movers can best be described as reciprocating steam engines driven instead by water-pressure. Small and compact, water-pressure engines were well suited to mine pumping (and also power generation in hilly country) for two reasons. In the first place they could use any head of water, consistent with piping and engine being able to withstand the applied pressure, and in this respect water-pressure engines were a marked advance over vertical and horizontal water-wheels whose capacity to utilize a large head was very limited. And secondly the failure of a water-pressure engine underground, although inconvenient and perhaps expensive, was not likely to be dangerous. By contrast the steam engine, especially in its high pressure form, represented a terrible liability if used in a mine. In one important sense the water-pressure engine was an anticipation of things to come. When the water turbine eventually established itself at the expense of all other hydro-power engines it depended on a feature already present in water-pressure engines, namely the use of a piped supply utilizing as much as possible, if not all, of the total available head of water; this common feature marks the two engines as a distinct advance over the old fashioned water-wheel and is indicative of a new approach to and quest for greater efficiency and higher power.

If the water-wheel did not figure prominently in mine pumping, the same is not true for certain other industries. In so far as it was at the time the only source of industrial power, the water-wheel must be regarded as an important factor in the formative decades of the Industrial Revolution. The textile and iron working industries in particular were heavily dependent on water-power and its availability strongly influenced their development and location. In recent years it has become increasingly clear how much the growth of industrial centres in Yorkshire, Lancashire, Cheshire and Derbyshire depended on the twin assets of rain and the Pennines. James Watt's rotative steam engine was first put to work (in John Wilkinson's ironworks) in 1783, but it by no means quickly – and in some places not even slowly – replaced the water-wheel, something which requires more than an appeal to tradition or conservatism for an explanation.[2]

A well made water-wheel with a properly regulated water-supply would run at a very steady rotational speed which for many a textile operation or iron working process was regarded as an advantage of real worth. Moreover, the sheer simplicity of water-wheels combined with a long-standing experience in their construction obviated any serious likelihood of breakdowns. And they were cheaper; cheaper to install in the first place, and even more so in operation because whereas water was free coal was not, and prior to 1800 Boulton and Watt vigorously exacted their famous premium from each and every customer.

Evidence that hydro-power resources were extensively developed is not hard to find. Richard Hills has detailed[3] numerous examples of the efforts made by the textile industry to harness as much hydro-power as could be found. Some rivers quickly reached the point where a very large proportion of the total available fall was being utilized. Since such developments also involved the construction of mill-ponds (so as to make as much use as possible of total flow as well as total fall), it is not surprising that many a stream became mightily congested, to the detriment of existing operations and future improvements. There were numerous disputes arising from full reservoirs flooding the tail races of higher mills and a variety of problems occurred when a new wheel was erected between two existing ones. A rough and ready code of behaviour gradually evolved to protect riparian interests and eventually this formed the basis of the formal legal code which developed in the nineteenth century. None of this, however, could obscure a fundamental problem. The utilization of a hydro-power resource can ultimately be exploited to the point beyond which no further developments are feasible. When all the fall and all the flow of a river have been harnessed, then short of massive engineering works of the sort that divert one river into another a limit has clearly been reached. Regularly in the eighteenth century and more frequently in the nineteenth this state of fluvial exhaustion was achieved. For instance, Baines writing in 1835 states[4] that on the River Irwell some 300 mills were making use of 800 feet of fall from a theoretical total of 900 feet.

The gradually developing shortage of economic water-power sites opened the way for the construction of steam engines. The impact of steam power was felt in two stages. The Newcomen engine and the early Watt engines were pumping engines, and factory owners were quick to see that one solution to their power problem was the installation of steam powered pumps at water-power sites. Thus water could be recirculated, an arrangement which seems first to have been used at Coalbrookdale in 1742 and in the 1760s proved to be a temporary solution to the water shortage problem which bedevilled the Carron Company near Falkirk. Subsequently the system enjoyed quite a vogue, so much so that it was possible for Joshua Wrigley to manufacture and sell a developed version of the original steam powered pump which Thomas Savery had patented in 1698. Direct drive of factory machinery by rotative steam engines did not really assert itself until after 1800, and even then water horsepower probably continued to exceed steam horsepower well into the nineteenth century.

The water-wheel experienced numerous constructional developments during the heyday of its use in industry. A natural progression was from wooden to iron construction, the peripheral parts and axle components making the change somewhat before the other parts. The all-iron wheel began to be used regularly in the nineteenth century. Immediately noticeable is a lighter-weight form of construction, frequently of very pleasing appearance, although this was not solely the result of all-iron construction. Partly it derived from the practice of taking the drive off the wheel's periphery, thereby relieving both spokes and axle of the need to carry anything like such large forces as before.

British conditions greatly favoured the construction of breast-wheels, so much so that such specialized terms as low, medium and high breast-wheel came into use. Another variant was the 'pitch-back wheel', a breast-wheel so high that it was all but an overshot-wheel, but with one distinction. Water was supplied just short of the top so that the direction of rotation was as in a breast or undershot wheel. The advantage was that water drained from the buckets in the same direction as the tail race (see Figure 34).

Famous names are associated with water-power developments in Industrial Revolution Britain. William Fairbairn for instance, who apart from his involvement in a number of hydro-power schemes, was responsible for the 'ventilated bucket', a simple yet elegant idea which allowed

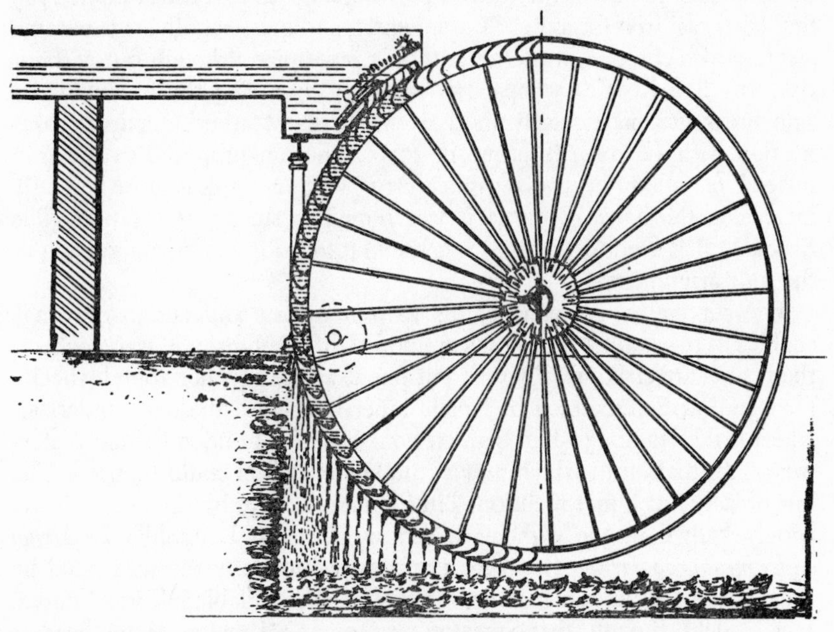

Figure 34 The pitchback water-wheel which was widely used in Great Britain in the nineteenth century; this one was built by Sir William Fairbairn.

air and water to change places quickly in the buckets at the moments of entry and exit of the water. John Rennie, likewise a builder of water-wheels, contributed his 'sliding hatch', a flow control device designed to sustain as high a supply level as possible and so maintain the maximum possible head of water. The major figure, however, was John Smeaton, the 'father of civil engineering', who throughout his varied and illustrious career was hardly ever without involvement in some water-power project or other.[5]

Smeaton's water-wheels were applied mostly to milling (mainly corn, sometimes oil, paper and even snuff), and this group includes some of his biggest wheels, as much as 30 feet in diameter. The most interesting group though are the 'engines' designed for the iron industry: six for working bellows and hammers at Kilnhurst forge; a pair, contra-rotating, for a rolling mill on the River Coquet; and most impressive of all, the blowing engines at Carron. Smeaton's association with the Carron Ironworks extended over more than twenty years – from 1766 to the end of his life – during which time he was responsible for improvements to and the design of water-wheels, blowing engines, boring machines and much other equipment in the works.

John Smeaton's long association with Carron convinced him more than anything else of the value of iron as a machine building material. It was for the wheel of one of the Carron blowing engines that Smeaton in 1769 first tried cast iron for axles. Subsequently, and quite rapidly, this became a standard technique in the millwright's repertoire although not without risk. In 1819 Rees' 'Cyclopaedia' was still discussing the possibility of axle breakages 'particularly in cold and frosty weather'. Again in connection with Carron, this time in 1770, Smeaton proposed to use iron instead of wood for the peripheral rings of an undershot wheel. By increasing the wheel's inertia it was hoped to steady its rotation. The Brook Mill at Deptford (1778) was the first to feature cast iron gearing in the mill machinery.

The survival of so many of his written reports and design drawings enables us to examine Smeaton's opinion of different forms of water-wheel; the results are striking There is but one example of a horizontal wheel – for a mill at Scremerston in Northumberland – and the only undershot wheel is the mighty 32-foot diameter one built for London Bridge Waterworks, a situation in which *only* an undershot wheel could be used. The low breast-wheel, first mentioned in England in 1539 by Fitzherbert[6] and widely known in the eighteenth century through Leupold's *Theatrum Machinarum molarium* (Leipzig, 1735), was frequently recommended by Smeaton as the best arrangement wherever the available fall was limited. It is evident, though, that Smeaton was convinced that as high a head as possible should be utilized. Consequently, overshot and pitch-back wheels figure prominently in the lists of his projected and completed works and this is in no way surprising. Quite early in his career Smeaton had conclusively demonstrated the superior efficiency of overshot wheels by means of a series of brilliantly conceived and executed experiments.[7]

In 1752, when he began his water-wheel experiments, Smeaton was twenty-seven years old and working in London as an instrument maker. Probably it was through his attendance at meetings of the Royal Society that Smeaton became interested in the question of water-wheel performance. The matter was very much in the air and a number of authorities had already been moved to pronounce on some aspect or another. Edmé Mariotte, for instance, in his famous *Traité du mouvement des eaux et des autres corps fluides* of 1686 had attempted to estimate the force exerted by a stream of water on a stationary plate. In 1704 Antoine Parent asserted that maximum power would be developed in an undershot wheel when the peripheral velocity was 1/3 the stream velocity; and the power developed, claimed Parent, could never exceed 4/27 of that potentially available in the water. Parent's analysis, notwithstanding its several and notable defects, was well known and widely quoted in the first half of the eighteenth century.[8] Both Stephen Switzer in *Hydrostatics and Hydraulics* (1729) and Bernard Forest de Belidor in his *Architecture Hydraulique* (1737), adopted Parent's view and also advanced the idea that undershot wheels were six times more efficient than overshot ones. J. T. Desaguliers, on the other hand, stated in *A Course in Experimental Philosophy* (1734) that overshot wheels were ten times as efficient as undershot ones. So, between Belidor and Switzer on the one hand and Desagulier on the other stood a factor of no less than sixty!

By the middle of the century, the desirability of resolving what Thomas Telford was later to term 'such monstrous disagreement' had become clear to Antoine Deparcieux in France and John Smeaton in England. Both, moreover, had a shrewd idea that 4/27 was much too low a figure for the efficiency of actual water-wheels. Deparcieux's work, which became well known in France, was based on theoretical considerations as well as experiments; John Smeaton depended solely on 'experiments made on working models, which I look upon as the best means of obtaining the outlines in mechanical enquiries'. Experiments to determine the efficiency of water-wheels had already been made by Christopher Polhem during the early part of the eighteenth century, but to Smeaton falls the distinction of being the first to test models, the earliest application in any field of engineering of scale model analysis. It was an exceptional piece of work. Particularly interesting is Smeaton's realization that, due to scale effects, relating the results of model tests to the performance of the full-scale article is by no means so obvious as it looks. Consequently, the results of the research of 1752–3 were not made public until 1759, by which time Smeaton had been able to make confirmatory observations of working wheels, some of which were his own designs.

Among the many important results of Smeaton's experiments, the following were of direct and special interest to hydro-power men.

1. Overshot wheels were about twice as efficient as undershot wheels; in modern terms the ratio was 66% efficient as against 30%.[9]
2. Overshot wheels were driven by weight of water alone and precious little was gained by allowing the applied stream of water to strike hard against the buckets.
3. For undershot wheels the best velocity ratio between wheel and stream was shown to be at least 2:5 and under certain conditions approached 1:2; both were less than Parent's 1:3.
4. Impact between a stream of water and a flat plate resulted in a marked loss of energy in the form of spray and turbulence; or, as

> Smeaton observed, 'nonelastic bodies, when acting by their impulse or collision, communicate only a part of their original power; the other part being spent in changing their figure in consequence of the stroke'.

Having established the superiority of overshot water-wheels – and therefore pitch-back wheels as well – Smeaton favoured their construction wherever conditions allowed. Otherwise he advocated breast-wheels, this type being discussed at the end of Part II of his paper where he proceeds 'to examine the effects when the impulse and weight are combined, as in the several kinds of breast-wheel'. Smeaton points out that even a very low breast-wheel will perform better than an undershot wheel because at least some use is made of the water's weight.

To what extent the paper of 1759 was known to Smeaton's English contemporaries and heeded by them is difficult to detail. It *is* clear though that towards the end of the eighteenth century undershot wheels were ceasing to be built. It is reasonable to suppose that this demise was initiated by Smeaton's research and later on sustained by the example of his own professional work. By the beginning of the nineteenth century engineers were already inclined to speak of the undershot wheel as if it were a historic machine to them.

Nineteenth-century Continental engineers make regular references to the work of Smeaton, the only British engineer whom they acknowledge in a field in which enquiries were numerous on the Continent in the eighteenth century, especially in France. It was in 1759 that Antoine Deparcieux published the results of his work on much the same lines as Smeaton's but including theoretical discussion as well as tests on models. Better known and more frequently quoted were the results obtained by Charles Bossut in the 1760s and published in 1771 in his two volume work, *Traité théorique et expérimental d'hydrodynamique*. Bossut was a devoted experimenter in hydraulics and, like Smeaton, was acutely aware of the dangers and difficulties involved in correlating results obtained from models with the real thing. He notes, moreover, the difficulty of choosing a size for the model, a large specimen being the more difficult to test, a small model being prone to scale effects.

Bossut's undershot water-wheel experiments, in addition to confirming the essence of Smeaton's results, also included some different conditions. He tried, for instance, to establish whether the wheel's performance was different in a closed flume as against a broad, open channel; the results suggested very little difference. On the other hand he noticed a marked difference in a wheel's performance if the number of blades was increased from 12 to 24, but that the improvement was proportionally very much less when the number was doubled again up to 48 blades. Bossut also found that inclining water-wheel blades between 15° and 30° away from a radial position was beneficial. In the so-called Sagebien wheel this modification was applied to a practical machine and in its all-iron form the type was widely used in the nineteenth century in France.

Both Smeaton and Bossut were much interested in the optimum velocity ratio of undershot wheel to stream. Their experimental results ranged from 1:3 to 1:2 and both of them settled for 2:5 as a reasonable compromise in most practical situations. This matter was to receive final clarification at the hands of Jean Charles Borda.

The Chevalier de Borda, whose varied and interesting career as military, hydraulic and naval engineer has by no means been adequately studied, applied his attention to hydraulics and hydro-machines in the 1760s. His *Mémoire sur les roues hydrauliques* was submitted to the Académie des Sciences in 1767 and published in 1770.[10]

In Parent's work of 1704 in which it is suggested that the velocity of a water-wheel blade should be one-third the stream velocity, the basic error lies in the assumption that an undershot wheel can be represented by a single blade propelled by a jet or stream of infinite length. Borda realized that, in fact, because a *series* of blades intercept the stream in rapid succession the problem is somewhat different, and he showed correctly that the peripheral speed of the wheel should be half the stream speed in order to extract maximum power from the water. This conclusion was largely confirmed by a number of the experimental instances investigated by Smeaton, Deparcieux and Bossut, especially in those tests where relatively large quantities of water at low heads were tried.

Although Borda had been anticipated in considering a number of facets of hydro-power problems, he was the first to assemble anything like a basic set of propositions along with practical equations with which to determine actual quantities. Already Daniel Bernoulli, John Desaguliers, Leonard Euler and John Smeaton had considered the question of the energy lost in an inelastic collision. Smeaton, it will be recalled, points out that a fundamental advantage of the overshot wheel is its independence of any impact effect. Borda went further and stated, as a basic hydro-power proposition, that water should be applied to the prime mover without percussion, the first clear reference to what is usually known today as 'shockless entry'.

The fact that water pouring down a mill-race represents a waste of hydraulic energy was another point to which Daniel Bernoulli first drew attention and some thirty years later it was Borda who formally stated that for maximum efficiency water should always leave the wheel without velocity.

Of course, given these conditions, one or two basic difficulties are immediately apparent. The idea that water can escape from a water-wheel 'without velocity' is absurd. Even more troublesome, in the case of undershot wheel, is the requirement that the blade velocity should be half the incoming stream velocity, because this means that water leaves the wheel with the same velocity as the blades and a very large proportion of the available hydraulic energy *has* to be run to waste even when the engine is yielding its maximum power. So an important aspect of Borda's analysis is his demonstration that the maximum theoretical efficiency of a hydraulic prime mover is not necessarily 100%. John Smeaton in a later paper of 1776 deduced the same thing, to judge from his statement that 'in large works the effect is still greater, approaching towards half, which seems to be the limit for the undershot mills, as the whole would be the limit for the overshot mills, if it were possible to set aside all friction, resistance from the air, etc.'[11] Smeaton's observations are essentially correct. The undershot wheel is at best 50% efficient, as Borda demonstrated, and it is in the nature of the hydro-dynamics of the machine that it can never be more. The overshot wheel on the other hand can, in principle, reach 100% efficiency.

Both Smeaton and Borda had a common interest in another aspect of overshot water-wheels, namely relative and absolute speeds of water-supply and buckets. One of Smeaton's basic conclusions from his experiments was that contemporary practice favoured wheel speeds which were much too high. Thus, depending on a wheel's size, he judged that peripheral speeds need only be in the range 2–3 feet per second. It was Borda's contention that the velocity of the water supply should be no more than double the wheel speed. Smeaton and Borda between them, then, showed that the overall operation of an overshot wheel could be a much more leisurely affair than was usual at the time. Subsequently, however, in actual designs for overshot wheels, engineers opted for speeds somewhat higher than theory predicted because in the words of Thomas Tredgold, 'the water-wheel which has a quick motion acts as a fly to regulate the motion of the machinery, whereas, if it move slowly, it requires regulation itself'. Generally speaking, wheel speeds in the range of 4–8 feet per second became usual in nineteenth-century overshot wheels.

The Chevalier de Borda's statement to the effect that for maximum performance water should enter a hydro-power machine without shock and leave without velocity was reiterated in 1783 by Lazare Carnot as follows: 'In order that a machine should produce all its effects, it would be necessary, first, that the fluid should absolutely lose all its motion by its action upon it; second, that it should lose all this motion by insensible degrees, and without any percussion.'[12] This was the most basic statement so far made on the theory of prime movers and we shall see now (and at a later point in this chapter) the extent to which these propositions were instrumental in shaping practical developments.

The eighteenth century was essentially a period of experiment and explanation. Improvements in vertical water-wheels did occur as a result, the career of John Smeaton being an especially well documented instance, but it is not until the nineteenth century that we encounter any conscious effort to realize in full the precepts contained in the Borda–Carnot statement of the ideal fluid machine. A well known and striking example is the work of J. V. Poncelet.

As a scientist, Poncelet was well able to appreciate the basic defects in the principle of the undershot water-wheel; at the same time, as a realist and a practical engineer, he knew full well that the machine was a relatively inexpensive proposition in terms of both initial and running costs. He therefore addressed himself to the idea that it might be possible to devise an undershot wheel which received water without shock and exhausted

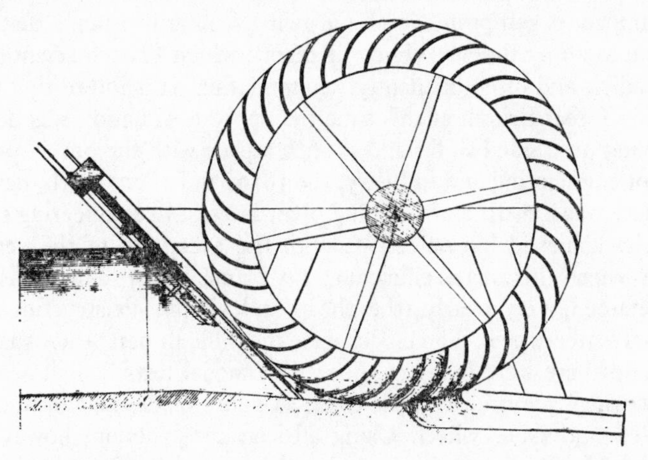

Figure 35 The Poncelet wheel as illustrated by its creator in 1827.

it without velocity, as theory required. Thereby he planned to combine high efficiency with low cost; the famous Poncelet wheel was the result.

As Figure 35 shows Poncelet realized his aims with little elaboration of the basic undershot layout. At the delivery side of the wheel he used a deep angled sluice to generate a broad jet whose direction coincided exactly with the leading edge of each curved blade as the latter moved down into a position to be struck by the jet. In this way, in all practical terms, shockless impingement of water on blade was achieved. As it 'climbed' the curved blade the water gradually gave up its energy. Running back along the blade would restore the water's velocity (always neglecting friction effects) *relative* to the blade. At exit from the wheel, however, the water's *absolute* velocity is reduced to zero. Poncelet was thus able to realize, at least on paper, the theoretical ideal.

To prove his designs Poncelet carried out experiments on a model and a full-sized prototype. He found that provided the correct ratio of wheel to water velocity was achieved – somewhere in the range 0·5–0·6 – then efficiencies of 60–70% could be expected. In practice this proved to be true only in situations where large quantities of water were delivered at low heads.

The theory of Poncelet's wheel and the results of the model tests were published in 1825 in a paper, *Mémoire sur les roves hydrauliques à aubes courbes mues par dessous.* Immediately it was awarded the Prix de Mécanique of the Académie Royale des Sciences, and subsequently in 1825 and 1826 appeared in several different journals. In 1827 it was published as a separate monograph.[13] This rapid and widespread circulation of Poncelet's work is of interest because no other paper on water-wheels, not even Smeaton's, had proved so interesting and commanded such attention. Of course Poncelet's paper is much later, nearly seventy years, but more than that its impact is symptomatic of a growing feeling in France that it was high time to apply the large body of theory which French scientists had been steadily, and often brilliantly, accumulating. It is interesting to note too that in Great Britain at this time the opposite situation was developing. Having proceeded so far and at such a pace with the purely practical aspects of engineering and industry, the 1810s and 1820s mark the period when there was a distinct awakening of an interest in engineering science.

Poncelet achieved his objective. With the exception of the very best overshot wheels, his was as efficient as any prime mover so far devised and yet it retained, very nearly, the cheap and simple construction of the undershot water-wheel. The blades were the difficult part; their shape was critical and there were many of them. The model tests had shown that a Poncelet wheel required at least twice as many blades as the traditional flat-bladed undershot wheel. Using all-iron construction, however, the fabrication of a Poncelet wheel was by no means difficult and many were built.

The rationally designed undershot water-wheel, for that is what Poncelet's wheel represents, was not the only notable result of the surge of interest in theoretical hydraulics and hydraulic motor design which was a feature of French engineering in the 1820s. Into the hydro-power engineering vocabulary of the time came a new word, *turbine*, and it is to the history of this machine that we must now turn.

13

Early Turbines

BEFORE EXAMINING THE complicated history of the two basic forms of modern turbine, the impulse turbine and the reaction turbine, it is necessary first of all to give brief definitions of the machines' modes of operation and then, as an aid to the enquiry, to change the nomenclature.

The so-called 'impulse turbine' (see Figure 36) works by first of all

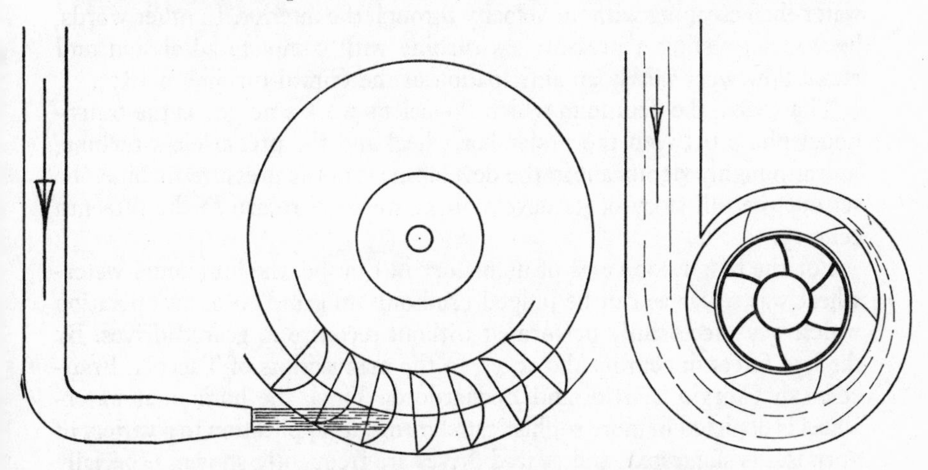

Figure 36 The basic forms of water turbine: *on the left* a pressureless turbine; *on the right* a pressure turbine of the inward flow variety.

converting all the available hydraulic energy into a jet (or jets) which is then played onto a wheel of buckets or chambers. At these buckets there is a large change of momentum but no change of pressure. In theory all of the kinetic energy in the jet is transferred to the wheel as mechanical energy. By contrast, the so-called 'reaction turbine' takes in water with velocity and also under pressure. As the water flows across the turbine's blades mechanical energy is produced by converting both the kinetic energy of the water *and* its pressure energy. At exit the water has a high relative velocity and its pressure is atmospheric (or less if a draft tube is fitted). Of course, both types conform to the principle of shockless entry and discharge the 'used' water with as little absolute velocity as possible.

The conventional nomenclature, impulse turbine and reaction turbine, is not particularly accurate because in fact both machines are driven by reaction. The real difference between them is that one operates at atmospheric pressure, all the available energy being in kinetic – i.e. jet – form, while the other is under pressure *at the wheel* and no jet is formed. It is perhaps more reasonable, therefore, to use the terms 'pressureless turbine' and 'pressure turbine'; and for a historical discussion it is much more helpful.

In the light of the above descriptions it can be observed straightaway that Poncelet's wheel is a very crude form of pressureless turbine in which a stream of water, rather than a properly formed jet, imparts its energy by change of momentum as it passes over a series of curved blades. The machine is otherwise imperfect in that water enters and leaves the wheel at the same point. Poncelet recognized this defect. In a lecture given at Metz in 1826, he suggested how the Poncelet wheel could be mounted on a vertical axle and supplied with water right round its periphery, the spent water then escaping without velocity through the interior. In other words, he was suggesting a pressureless turbine with complete admission and radial flow, very nearly an anticipation of the Girard turbine of 1855.

The 1820s, the decade in which Poncelet's work emerges as the transitional phase between the undershot wheel and the pressureless turbine, was also highly significant in the development of the pressure turbine. To set out the full story of its development we must return to the fifteenth century.

For the first 1,000 years of its history in Europe, the horizontal water-wheel was so far as can be judged used only to grind corn, an operation which was successfully performed without recourse to geared drives. By the late fifteenth century, however, in the manuscripts of Taccola, Francesco di Giorgio Martini and Leonardo da Vinci, the horizontal water-wheel is depicted in more sophisticated forms, its application to a variety of purposes is suggested, and geared drives are frequently shown, especially in Martini.[1]

A century later, in Juanelo Turriano's *Los veinte y un libros de los yngenios* (1569), the horizontal water-wheel continues to predominate in the many water-powered installations depicted.[2] Prominent, of course, is corn grinding of the 'gearless' variety but in addition there is a wider range of other operations than had hitherto been covered. Roller mills, burnishing and grinding wheels, and devices to cut and crush sugar cane are all shown. A water-powered sugar cane plant is illustrated with a geared drive to a horizontal shaft and there are variants of this arrangement. One such shows the drive being taken straight off the water-wheel itself by means of a set of peripheral teeth which are very nearly extensions of the blades themselves.

Although ideas for extending the application of the horizontal water-wheel found frequent and very often ingenious expression on paper any substantial evidence that such developments took place in practical engineering is not forthcoming, at least not yet. The horizontal water-wheel began life as a gearless corn grinder and that was its destiny. In a recent and splendid article[3] on the survival of the horizontal wheel in south-eastern Europe, Louis C. Hunter has gone so far as to adopt the terms grist-wheel and grist-mill to signify (in those regions where it is still used) the almost exclusive role of the horizontal wheel right down to modern times.

In Europe, the use of horizontal wheels has declined steadily in the last century. Northern Europe abandoned the machine first, and in the valleys of the northern Alps in France, Switzerland, Germany and Austria – regions whose topography and hydrology made it a very attractive proposition at one time – a decline seems to have set in during the nineteenth century. By comparison, the Mediterranean slopes of the Alps and also many Pyrenean valleys continued to use horizontal wheels in some numbers until twenty years ago; but working examples are very hard to locate now.

In the south of Spain the horizontal wheel has had a long history of use; Plate 9 shows a tiny scale model of the type which until quite recently was grinding corn in the tenth-century mills at Cordoba. Further west in Iberia, Portuguese communities in small and remote places continue to use the grist-mill.

In Europe it is in the south-eastern region that horizontal wheels seem to have lingered the longest. Dr Hunter's admirable researches in the Balkans have indicated what a large number are still used in the wet and mountainous parts of Rumania, Bulgaria, Yugoslavia and Greece. Nowhere else is it possible to find such concentrations of them, sometimes a dozen or more crowded together at the one site if conditions allow.

The horizontal water-wheel's restricted area of application – to flour milling – and its incapacity to challenge the vertical wheel as a source of industrial power, have frequently been used to support the argument that the machine was an inherently inefficient prime mover. This view is not altogether satisfactory. After all, the efficiency of the horizontal wheel, say 10–15%, is not so much worse than the undershot which for many centuries was the predominant hydro-power engine in the West. The view that a vertical wheel driving a horizontal shaft is a more satisfactory configuration for most industrial purposes is not given weight by the water-wheel's history in China, where horizontal wheels were successfully used for such heavy work as furnace blowing and trip hammering.[4]

Nevertheless, and leaving aside the question of mere efficiency, it *is* the case that the vertical water-wheel exerted a great influence on the development of European industry and manufacture; most studies of Western technology emphasize the machine's history for that reason. However, from our point of view, in the matter of hydraulic prime mover concepts and variety of ideas, as opposed to questions of usefulness and extent of application, the development of the horizontal configuration is full of interest and in the end the most important results.

The late fifteenth-century manuscript drawings of horizontal water-wheels – notably those of Francesco di Giorgio Martini and Leonardo da Vinci – show the embryo pressureless turbine. Two essential components are present. First, there is the jet. Of course, horizontal wheels were frequently supplied by an open channel or chute, a practice which has predominated to the present day, but in Martini and Leonardo the jet makes its earliest appearance so far discovered. Details are difficult to elucidate from the somewhat imprecise drawings, especially Leonardo's, but in Martini some crudely fashioned square-section trunking is shown tapering down to a form of nozzle.[5] The execution may be faulty but the idea is notable. Over a century later, in Jacob de Strada's *Künstlicher Abriss Allerhand Wasser*, *Wind*, *Ross und Handmühlen* of 1617, much of which is evidently a wholesale borrowing of Martini's treatise, the trunking and nozzle are improved, not to mention the draughtsmanship.

Secondly, there is the shape of the buckets – or vanes, paddles, blades, whichever term happens to be conventional or appropriate. Beginning with the manuscript illustrations of Taccola, Martini and Leonardo, specially shaped buckets are shown, more often than not with double curvature. In these early instances the curvature is predominantly about a vertical axis so that the change of momentum of the jet takes place largely in a horizontal plane. This family of wheels, continued in the drawings of Ramelli, Strada, Belidor and others, has a record of regular use down to modern times.

Much less like a wheel of buckets and more like a propeller is the horizontal wheel first shown by Juanelo Turriano[6] and destined to evolve into a distinct type in its own right. It turns up in Belidor, while even earlier in Strada there is a heavily curved, nine-bladed wheel of very advanced appearance. This second group of horizontal wheels features a much greater component of axial flow than the previous type.

The horizontal wheels of the late fifteenth century and their derivatives – the lines of descent can be traced to the nineteenth century – are crude pressureless turbines; all the available pressure is converted to kinetic energy before the water meets the wheel, while the force exerted against the buckets or blades is achieved by passage of the water across curved surfaces. At the time ideas about shockless entry and zero exit velocity were completely absent and yet a process of evolution produced buckets and blades whose shapes are tending towards those which rational design would require. The first word on the theory of pressureless turbines can be found in Borda (1767), where the ever perceptive Jean-Charles examines the requirements of shockless entry and zero exit velocity in a double shrouded horizontal wheel fed by an open chute.

It is most interesting that empiricism made much greater progress in developing the horizontal wheel than the vertical. As we have seen, virtually nothing happened in the latter's case until the mid eighteenth century, and not until the nineteenth did rational design equip the undershot wheel with curved blades. Yet 300 years earlier curved blades, and not infrequently jet feeds, were features of horizontal wheels. It remains a little paradoxical that the big water-wheel which powered industry was a huge brute of an engine, very unrefined and little understood until its very last phase, whereas the horizontal wheel, essentially a low-powered engine limited to simple rural work (it is often spoken of as the primitive horizontal water-wheel), was, in a hydrodynamic sense, relatively advanced.

From quite an early date horizontal wheels were frequently shrouded, that is to say, the tips of the blades or outer edges of the buckets were attached to a cylindrical casing. This arrangement would help to strengthen the whole construction and also, and perhaps more important, served to eliminate or reduce splashing; thereby the wheel's performance was bound to be improved. That splashing and spray was detrimental to the wheel's performance must have been manifest, and even before 1600 in an effort to solve this problem horizontal wheels were made to turn inside close fitting cylindrical chambers. They are often known by their French name *roue à cuve* which is usually translated into English as 'tub-wheel'.

Juanelo Turriano's manuscript discusses tub-wheels, although he does not use that name, and recently these machines 'starred' in a lengthy discussion about the efficiency of horizontal water-wheels and the question of the turbine's origins.[7] Turriano deals with two sorts of wheel; the *molino de bomba* and the *molino de regolfo*. The *bomba* wheel is manifestly a jet wheel featuring twenty or more spiral blades, sometimes shrouded but not always, and fed from a tapered penstock acting under the full pressure of the water supply. The *regolfo* wheel (the name, significantly, means vortex wheel) is somewhat different and features only five or six blades of greater curvature.

It has been suggested that Turriano's *regolfo* wheel is a proto-form of pressure turbine but this is a notion to be approached with some caution. As Figure 37 shows, the *regolfo* wheel retains the characteristic small bore, high pressure, high velocity jet and enclosure in a chamber need represent nothing more than an effort to contain this jet. Nor is the depth of the chamber necessarily indicative of any attempt to induce a deep vortex, or for that matter a shallow one. In a mill designed to utilize a high head, 20 feet or so, structural reasons for a deep chamber are just as persuasive. In the absence of any performance data for *regolfo* wheels (and *bomba* wheels or any other early horizontal water-wheels for that matter), the late Dr L. Reti's bold claim that 'the modern counterpart of the *regolfo* mill is not the Pelton wheel but the Kaplan turbine'[8] is too simple, and in any case overlooks countless increments of development in the centuries between. A significant step was to do away with the jet altogether.

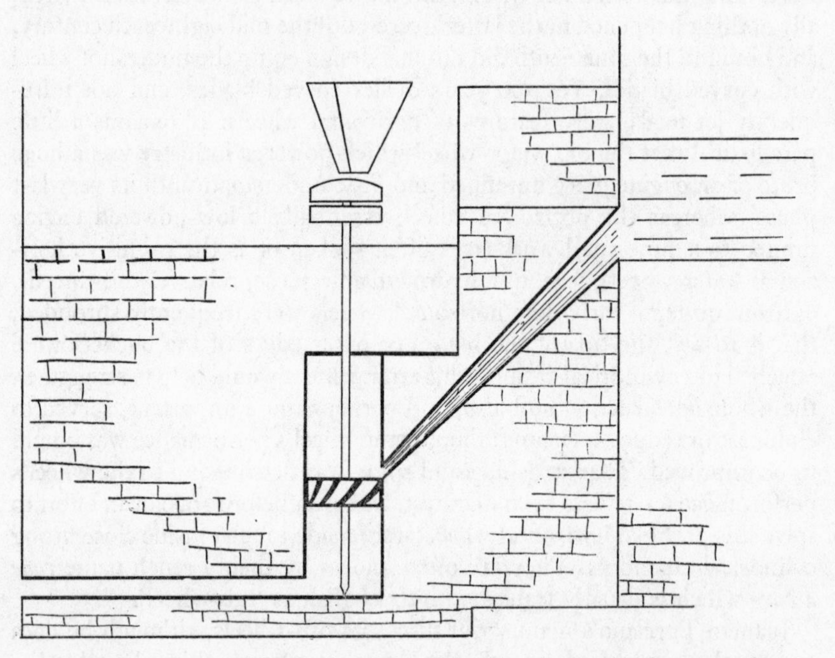

Figure 37 The regolfo wheel re-drawn from Juanelo Turriano's manuscript.

According to Buchetti's *Les Moteurs Hydrauliques Actuels* (1892), a horizontal water-wheel which had successfully dispensed with the jet was in use in southern France by 1620. This machine comprised an axial flow runner enclosed in the base of a cylindrical chamber into one side of which a deep tapered sluice poured large quantities of water. Buchetti's example from the Département du Gard is shown in Figure 38. D'Aubuisson says[9] that the same type of wheel was frequently used to grind corn on the Garonne, the Aude, the Tarn, the Aveyron, the Lot and other rivers where there is a large flow but relatively little fall. The jetless *roue à cuve* was a product of these conditions, especially at Basacle.

The horizontal water-wheels at Basacle in Toulouse on the River Garonne – they are often referred to as 'Belidor's wheels'[10] – appear to have been installed about 1700. One is shown in Figure 38 and its close resemblance to the Gard wheels is evident. The wheels of Basacle have dimensions apparently typical of *roues à cuve* generally; a wheel of 1 metre

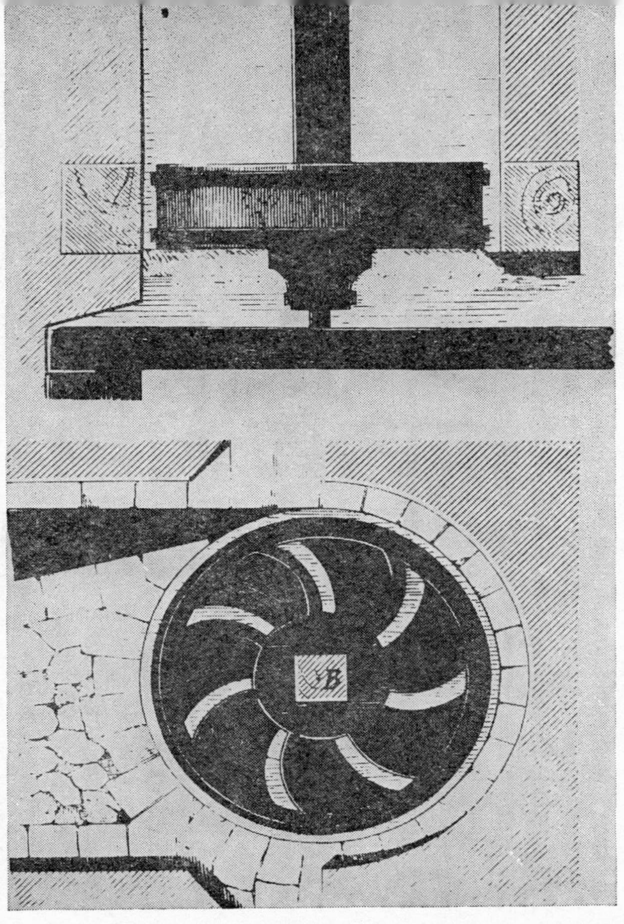

Figure 38 Two versions of the tub-wheel, the proto-pressure turbine: *on the left* a seventeenth-century version as illustrated by Buchetti; *on the right* one of the Basacle wheels.

diameter driven from a head of about 2 metres. As with the Gard wheels the tapered inlet sluice of Belidor's wheels would supply large amounts of water tangentially to the wheel chambers and with considerable rotational velocity. This vortex (note Turriano's term *regolfo*) would drive the wheel with a mixture of pressure and kinetic energy.

Belidor thought highly of the Basacle wheels and in some respects this was justified. They were straightforward and cheap to build (allowing that the curved blades could be made) and robust and convenient in operation. And provided that there was lots of water available, their poor efficiency was not a serious drawback. Belidor's approval has been taken by some as an indication that the Basacle wheels were not only proto-turbines in principle but also in performance, that they achieved the very high efficiencies commonly associated with modern machines. Belidor, though, was not always a good judge of machines. It was he, remember, who claimed that undershot wheels were six times as efficient as overshot, not the sort of statement to command confidence. When MM. Tardy and Piobert, artillery officers stationed in Toulouse, tested a selection of local tub-wheels on an early version of Prony's brake they found efficiencies for the Basacle wheels of only 12½–15%.[11] So although the tub-wheels of Provence, Languedoc and Gascony must be acknowledged as a notable advance towards the pressure turbine *idea*, the notion that they were fully fledged turbines in terms of performance is unrealistic.

In contrast to the Basacle wheels, a culmination of many centuries of development of the traditional horizontal water-wheel, the eighteenth century saw the appearance of a decidedly novel form of hydraulic motor which in the nineteenth century achieved a degree of useful application but only in a low-powered and inefficient form.

In J. T. Desagulier's *A Course of Experimental Philosophy* of 1734 appears the curious twin-jet device attributed to 'the learned and ingenious Dr Barker' and illustrated in Figure 39. In 1747 Professor Andreas Segner of Göttingen described his own developed version of the idea, using six jets instead of two. Segner's wheel, alike in principle to a lawn sprinkler, attracted the interest, among others, of Leonard Euler and his son Albert, who produced the very elaborate version shown in Figure 40.

One feature in particular of the Eulers' machine is notable. The lower conical drum, B, which is rotated by the efflux of water from the jets, b, is supplied with water at the top through a fixed cylindrical tank. So that water would enter the jet tubes smoothly (Leonard Euler was conscious of the need for shockless entry[12]), the feed tubes from the supply tank are appropriately angled. This is the earliest example of any hydraulic motor in which the inflowing water is properly directed to the full periphery of the runner. In a development of this arrangement the guide tubes are replaced by curved plates and here one has something which anticipates the guide vane system characteristic of modern turbines. D'Aubuisson claims that Eulerian turbines, although unwieldy and inefficient, were used in France.

The principle of Barker's mill continued to be exploited to the middle of the nineteenth century.[13] A typical version was the so-called Scotch turbine of James Whitelaw. Whitelaw, a turbine pioneer in Great Britain, made great claims for his product (probably unjustified) and by energetically canvassing for customers was able to sell them in the face of stern competition from the now well-established steam engine industry. Across

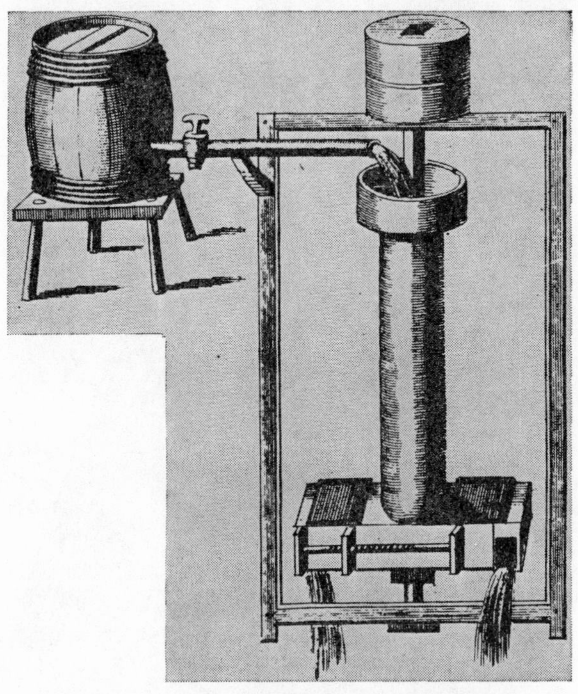

Figure 39 Barker's mill. (From J. T. Desagulier's *A Course of Experimental Philosophy*)

the Atlantic, a very similar reaction wheel was patented in 1830 by Calvin Wing and for a time in the 1840s there was a good deal of argument as to which was the better machine. In fact the question was hardly worth asking in view of the important advances being made with other and superior concepts.

Some general remarks can be made at this point. Prior to the 1820s it is possible to discern the pre-history of both the pressureless turbine and the pressure turbine. The embryo pressureless turbine is to be recognized in those horizontal water-wheels which utilized curved blades or shaped buckets and some form of jet. The pressure turbine's origins are more faltering and haphazard. The *roue à cuve* was widely used in certain parts of France and Spain but was very inefficient. Barker's mill claimed a degree of interest in the eighteenth century, was used in the nineteenth century, but never formed an important class of turbine. The Eulerian 'wheel' is of great theoretical interest but was little used as a consequence of numerous practical difficulties.

Just as the earliest concept of the turbine had taken shape in continental

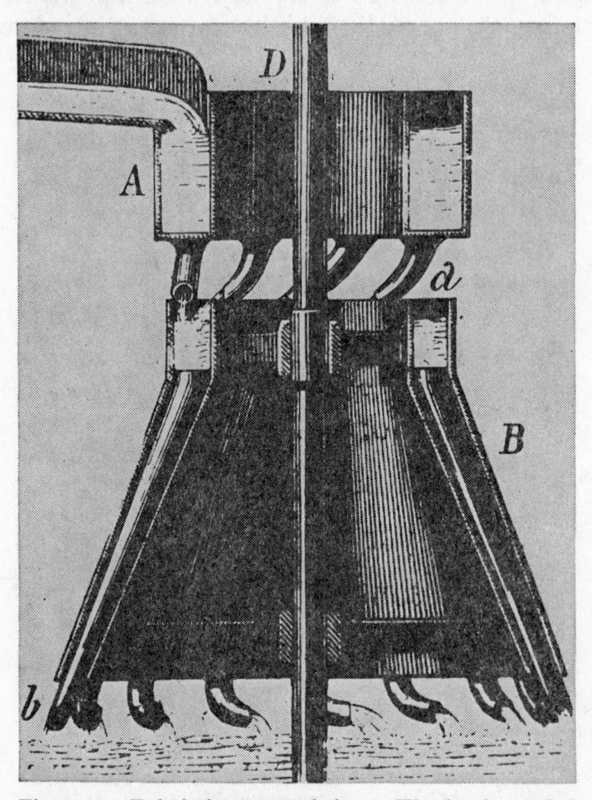

Figure 40 Euler's jet-powered drum. The header tank A supplies water through tubes to the revolving conical drum B which is rotated by the efflux of water from the jet tubes b.

Europe, so did the machine's formal development begin there and French turbine makers achieved the first commercial successes. As remarked earlier, in connection with Poncelet's work, early nineteenth-century France was in a mood to apply engineering theory to real problems and one problem in particular was crucial. Britain's Industrial Revolution – the very fact that the phenomenon occurred in Britain at all – was influenced in no small measure by the ease with which coal could be dug and the close proximity of this natural resource to another – iron. The development, and eventual dominance, of steam power was more easily accomplished in Britain than anywhere else. For very many European countries burning coal to create power for factories was not feasible, certainly not on the British scale, and it is symptomatic of France's lesser steam raising capacity that some of the English developments which made the steam engine more efficient were adopted more rapidly in France than in England.[14]

In order to develop and sustain industrialization, the French were bound to attempt the economic exploitation of the country's great reserves of hydro-power. Poncelet's wheel was rapidly adopted, and in 1826 the Société d'Encouragement pour l'Industrie Nationale offered a prize of 6,000 francs to anyone 'who would apply on a large scale, in a satisfactory manner, in factories and manufacturing works, the water turbines or wheels with curved blades of Belidor'. This competition was destined to run for a decade before a suitable winner was found.

An unsuccessful entrant for the prize in 1827 was Claude Burdin, an engineer of the Royal Corps of Mines and a teacher at the École des Mines in Saint-Étienne. The memoir which Burdin submitted covers a number of machines and the term *turbine* is used for the first time.[15] The most original and important of Burdin's proposals is shown in his drawing, reproduced in Figure 41, for an outward-flow radial turbine driven by the free efflux of water from a centrifugal runner. The stationary guide vanes and the runner blades are all shaped and angled in an effort to provide shockless entry and zero exit velocity.

Burdin's outward-flow turbine is historically significant only as a concept; practical application was another matter. Mere sketches and analysis were not enough to prove the machine's performance and Burdin never demonstrated so much as a working model, let alone a prototype. Moreover, it is clear from the drawings that contrary to some claims Burdin had not conjured up a pressure turbine in the modern sense; the lack of a casing and the considerable clearance between guide vanes and runner would have precluded any significant component of pressurized operation.

Burdin is hard to assess. Evidently he was not unfamiliar with the hydro-power theory and practice of the time, such as it was, and his written work displays some promising new ideas. That lack of perseverance prevented him from translating his concepts into working installations is a criticism which cannot be tested. At a practical level all Burdin could do was copy other people. This is evident from a sequence of contributions to *Annales des Mines* (1828, 1833 and 1836) in which Burdin describes some hydraulic motors which he set up in Puy-de-Dôme soon after 1820 (and prior to his 1827 memoir). All these machines, which Burdin rather grandly refers to as turbines, turn out to be variations on Euler's jet-propelled drum, evidently quite practical propositions but unexceptional in terms of power, speed, efficiency or new ideas.

Generally, it is the view that Burdin's principal contribution to the development of turbines was through his role as educator of Benoît

Fourneyron; and it is true that for two years Fourneyron was a pupil at the École des Mines and for a time in 1818 was taught by Burdin. After he had left the École in 1819 Fourneyron was involved in a wide range of engineering and industrial work, very largely in the fields of mining and metallurgy, the technologies which predominated around Saint-Étienne.[16]

In 1822 at Pont-sur-l'Ognon, Fourneyron installed a water-powered

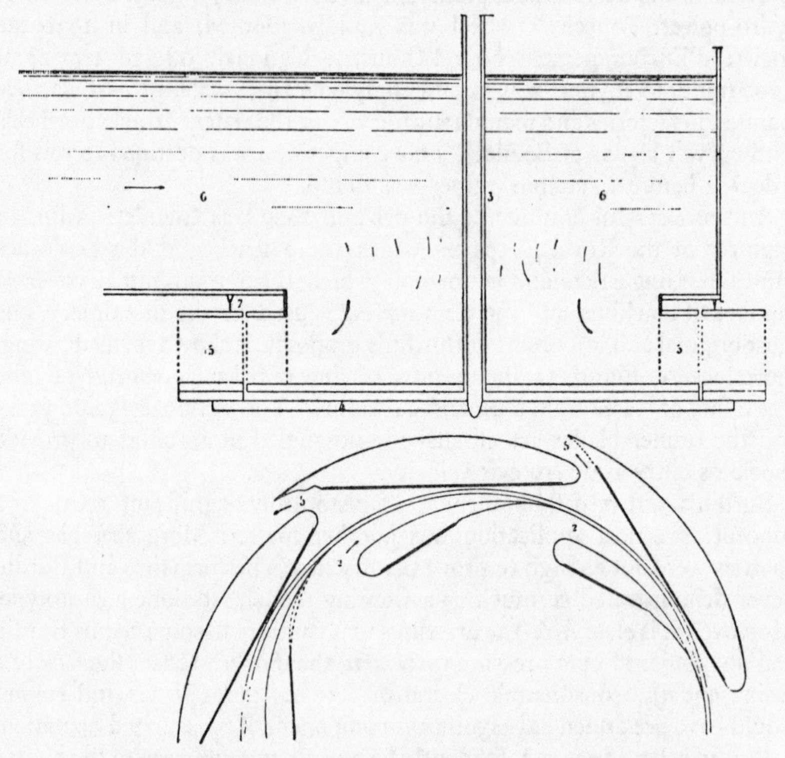

Figure 41 The first hydraulic prime-mover to be called 'turbine', Burdin's original six-bladed outward flow concept.

rolling mill in an ironworks and its highly successful operation prompted him to undertake more detailed studies of water-wheels and turbines. He began to study the written works of Euler, Bernoulli, Borda and Navier and of course the ideas of his own teacher, Burdin. It seems, therefore, that it was his own professional work that prompted Fourneyron's first interest in turbines, rather than his having been Burdin's pupil.

The first experimental turbine was developed and tested at Pont-sur-l'Ognon between 1823 and 1827. It was a small, low head (1·4 mts.) machine which could be run either unsubmerged or submerged. At its maximum efficiency, over 80%, it developed about 6 h.p. at 60 r.p.m. Like every other turbine which Fourneyron built, the Pont-sur-l'Ognon

Figure 42 This picture from Armengaud's *Traité des Moteurs Hydrauliques* shows a typical version of Fourneyron's early outward-flow radial turbine.

prototype was an outward-flow radial turbine with complete admission (see Figure 42).

By 1832 Fourneyron had developed his ideas considerably in the Dampierre and Fraisans turbines, both installed to drive blowing engines. In 1833 he entered a detailed account of the theoretical and actual performance of his three turbines for the 6,000 franc prize of the Société d'Encouragement. Fourneyron won. Belidor's wheels had at long last – after nearly a century – been 'officially' bettered.

Between 1832 and 1867, the year of his death, Fourneyron designed and installed well over one hundred turbines in France and other European countries. A few of his turbines went to the United States, the first about 1843, and their influence there was to be of some moment, as we shall see. In the case of a turbine ordered from Mexico, it was necessary to manufacture the whole machine in small enough component parts to be transported over mountainous country on the backs of mules.

Two of Fourneyron's turbines are of exceptional interest and a portent of things to come. In 1837 he provided two motors for a spinning mill at Saint-Blaise in the Black Forest; one operated under a head of 108 metres, the other 114 metres. These were heads of water of unprecedented size which not even the most advanced water-pressure engines had attempted to utilize. Needless to say, the development of a turbine for such extreme conditions confronted Fourneyron with some difficult problems. The design of guide vanes and runner blading appropriate to a pressure of 160 pounds per square inch (11 atmospheres) was unprecedented and so was the construction of sheet-metal penstocks capable of containing such high pressures. But ultimately the critical problem proved to be the bottom bearing of the turbine shaft which was so heavily loaded that it had to be renewed *every 10–14 days*.

The Saint-Blaise wheels' specification is a revelation, for these motors ran at 2,300 r.p.m., developed 60 h.p. at an efficiency of better than 80% and yet the wheels themselves were a mere 12½ inches in diameter and weighed 40 lbs. Actually in drawings of these installations, the runners are so small that they are by no means easy to find.

Unfortunately, want of space prevents a full analysis of many of Fourneyron's ideas, projects and achievements: his patent for a diffuser in 1855 for instance, his schemes to harness the power of the Rhine and the Seine and his various two-, three- and even four-stage turbines, a few examples of which were built.[17] We must, however, attempt to assess the position of the man who is possibly the best known of *all* those engineers who have contributed to the evolution of the water turbine.

Fourneyron's ability as a practical engineer is manifest. To have been so consistently successful in many branches of engineering, not just the field of turbines, is evidence of the surest of touches with machinery and materials. By comparison, it seems that Fourneyron was not a first-class theorist, as Poncelet was to observe in 1840. His reading of existing authorities, such as Borda and Burdin, was the principal content of his theoretical knowledge and he added nothing to it. Turbine design for Fourneyron was a matter of paying careful attention to the few basic principles which had been acknowledged and applying these to a turbine configuration which he, as a gifted engineer, sensed to be correct. The radial-flow concept applied with complete admission to a horizontally mounted runner was the best arrangement so far tried; and it concurred with the proposals of both Burdin and Poncelet.

Fourneyron's turbine, because it was such a step forward and was built in relatively large numbers, was frequently the subject of searching tests. For these studies Prony's brake was universally used and from the results, together with careful scrutiny of the turbines themselves, there emerge one or two basic weaknesses in Fourneyron's designs.

The performance of his turbines was impressive only under restricted conditions. At full gate, designed head and running submerged, high efficiencies were obtainable. But change these conditions – shutting down to part gate in particular – and the efficiency dropped sharply. Fourneyron's flow control mechanism, a cylindrical gate sliding up and down between the guide vanes and the runner, was effective as such but had one serious drawback. The wheel ran full only at full flow. At part flow some proportion of the wheel passage area was closed on the inlet side, the wheel could not run full, and there was, therefore, a sharp loss of power due to the sudden enlargement of the flow passage. Fourneyron was, of course, aware of this weakness which he attempted to deal with, fitting partitions into the runner to divide it into several 'storeys'. But this was only a partial solution. It is incidentally rather interesting to note that the specific speeds of Fourneyron's turbines were, with rare exceptions, so low that it is not correct to regard them as true pressure turbines. Fourneyron's pioneer machines, without a proper casing and with considerable clearance between guides and runner, were very largely pressureless turbines and not even submerged running or the 1855 draft tube were sufficient to change this materially.

The traditional view that accepts Benoît Fourneyron as the man who single-handed swept away all old-fashioned hydraulic motors and replaced them with the fully fledged pressure turbine is certainly over-simplified. What should be said of Fourneyron, and emphasized, is that he had some good basic ideas, made them work and embodied them in more than a hundred turbines which were sold to customers in several countries. It was through these highly successful working installations that Fourneyron put the turbine on the map.

A number of engineers, mostly French, such as Cadiat (1839), Callon (1840) and Huot (1852), attempted to develop Fourneyron's outward-flow concept.[18] They were not successful. If anything, they compiled confirmation of the defects already mentioned, while an accumulation of operating experience showed how serious was another problem, the tendency of an outward-flow turbine to race, sometimes to destruction, if there was a sudden loss or reduction of load.

The Fourneyron turbine was outmoded with almost uncharitable rapidity. The very success of the machine was a signal for a bewildering variety of ideas and innumerable patents all over Europe and North

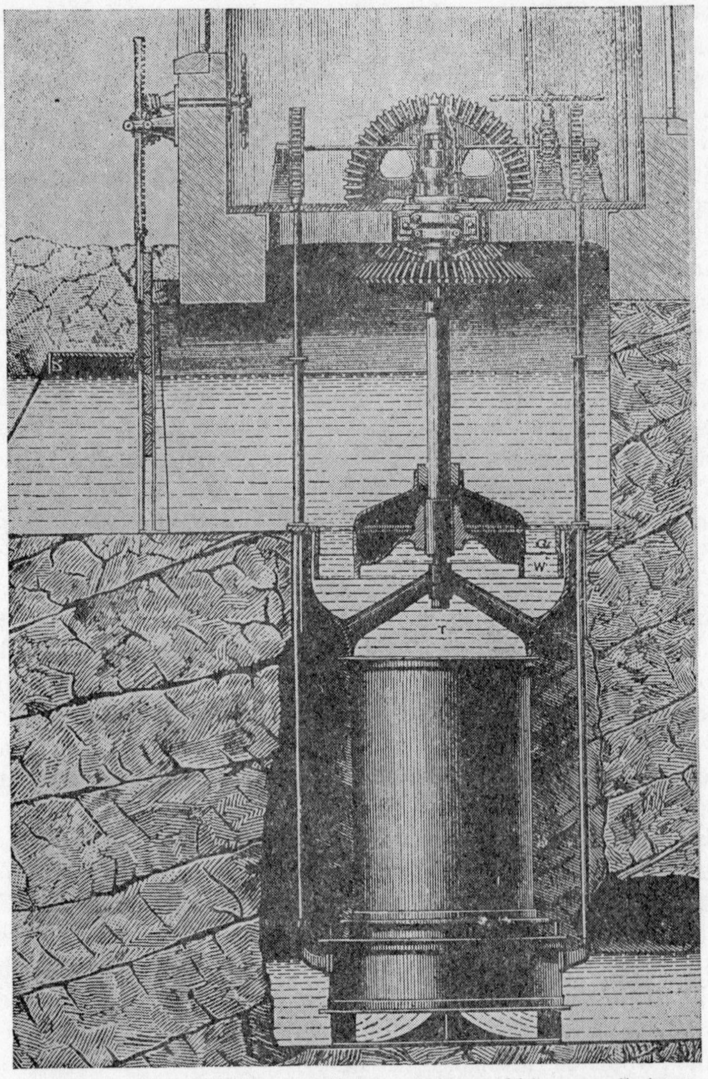

Figure 43 Jonval's pioneer axial-flow turbine could be set well above tail-water level by using a cylindrical draft tube.

America. Not until the end of the nineteenth century did this outburst of inventive activity and commercial rivalry begin to settle down and there then emerged a pattern of superior designs whose names are still famous. From a vast amount of nineteenth-century activity on both sides of the Atlantic we must try to pick out the key themes.

A trio of French engineers – Fontaine-Baron, Jonval and Koechlin – developed the axial-flow turbine.[19] An important feature of the Jonval turbine (1841), greatly improved by Koechlin in 1843, was its use of a draft tube, an integral component of the design from the outset. The Jonval–Koechlin turbine, shown in Figure 43, quickly established itself as the most suitable for heads of 3–5 metres while the overall simplicity and adaptability of the arrangement, especially the draft tube which allowed the runner to be located well above tail-water level, rendered it very popular. It was, moreover, a machine capable of quite high efficiencies at part gate, which in no small measure accounts for its lasting success. There is no reason to doubt Bodmer's claim (in 1889)[20] that it was the most widely used pressure turbine in Europe.

The designation pressure turbine can be applied with confidence to the Jonval–Koechlin machine. Whereas in the earlier turbines of Burdin, Fourneyron, Fontaine-Baron and others, no proper distinction between pressurized and pressureless operation was achieved, either conceptually or actually, the setting of the axial-flow turbine of Jonval precluded any component of pressureless operation.

Let us look finally, and briefly, at the European pressureless turbine. Following Poncelet's crude anticipation of the type and his suggestion for an inward-flow wheel with complete admission, many engineers in France, Switzerland and Germany showed a good deal of interest in Poncelet's ideas. Zupinger in Switzerland produced a single jet inward-flow type and Schwamkrug in Germany an outward-flow arrangment applied to a vertical wheel. Both seem to have been quite well known and performed satisfactorily. A tremendous variety of ideas characterizes the work of L. D. Girard, a French engineer who at one time or another tried both horizontal and vertical pressureless wheels, inward and outward radial-flow, axial-flow, part and complete admission.[21]

Girard's most lasting contribution was his outward radial-flow pressureless turbine, with either horizontal or vertical shafts and partial admission (see Plate 10). It was the best pressureless turbine of European origin and at the end of the nineteenth century some impressive units were in operation, sometimes as much as 1,000 h.p. being realized from a single wheel using heads of nearly 2,000 feet. The Girard turbine played a prominent role in the early development of European hydro-electric power.

Although the Girard turbine was easily the best of the European pressureless wheels it was not to find a permanent place in the turbine engineer's repertoire. Eventually it had to give best to its American rival, the Pelton wheel, one of a number of key nineteenth-century American developments whose history we must now consider.

14

The Turbine in America

EARLY UTILIZATION OF hydro-power in North America had two separate origins. In the south-west region Spanish influence introduced a whole range of hydro-technical expertise which in those arid lands found its principal outlet in irrigation. Water-power sites seem to have been developed only as an adjunct to irrigation schemes and details of what was involved are uncertain.

In the north-east things were very different. Northern European settlers in the New England states first began to utilize the area's rich hydro-power resources early in the seventeenth century as soon as they arrived.[1] The wheels themselves must have been the same simple vertical and horizontal varieties of the settlers' homelands and they were employed for those tasks which one would expect to be of paramount importance: in saw-mills, grist mills and powder mills. In a few places another early introduction was tide-mills, probably nothing very elaborate, but symptomatic of the settlers' need for mechanical power. As the frontier began to move westward so the water-mill moved steadily across the continent. These early water-power sites were effective in determining the patterns of settlement. Where water-mills were set up then people tended to gather and so towns grew up around the grist and saw-mills.

In the United States, as in Europe, the early decades of the nineteenth-century coincided with an increasing demand for power which was met in the main by breast- and pitch-back wheels, 10–20 feet in diameter, slow moving, made of cast iron, and amenable to increases of power by extending their length. The great majority of these wheels were the property of manufacturing companies who possessed the rights to harness the 'water powers' of New England's rivers. One such, and the most famous, was the Merrimack Manufacturing Company, situated at Lowell, the great textile making centre in Massachusetts. Lowell set the fashion for many aspects of New England's hydro-power evolution, including the employment of an 'Engineer of the Corporations at Lowell'. This was the post occupied by James Bicheno Francis whose book *Lowell Hydraulic Experiments* (1855) is one of the classics of hydraulic science.

Francis himself, while confirming the superiority and importance of breast- and pitch-back wheels in manufacturing, also draws attention to the widespread use of cheap, low-powered, horizontal wheels for milling. Their cheapness is emphasized: 'and in a country where water power is so much more abundant than capital, the economy of money was generally of greater importance than the saving of water.'[2]

By the middle of the nineteenth century American millwrights had made hundreds of attempts to develop improved water-wheels, Francis himself claiming that the United States government had granted some 300 patents for 'reaction wheels' alone. A few of these were moderately successful, such as Benjamin Tyler's 'Wry Fly' wheel, basically a *roue à cuve*. In 1838 Samuel B. Howd of Geneva, N.Y., made an important contribution in the shape of an inward-flow radial turbine. But for some reason he became dissatisfied with this arrangement and by 1842 was claiming all sorts of advantages for outward-flow machines. Much has been claimed in some quarters for Howd's influence, but really he remains an enigma. Probably it was what other people made of his ideas rather than what Howd did himself which was ultimately important.[3]

The early years of the 1840s mark the introduction into the United States of European ideas and European machines. Some examples of Fourneyron's turbine were imported and by 1843 a number of articles on its mode of operation and great commercial success had been published, notably in the *Journal of the Franklin Institute*.

In 1844 Uriah Boyden, a prodigious talent in hydro-power engineering, designed a Fourneyron-like wheel for the Appleton Company of Lowell, to be followed soon after by three more, each of 190 h.p. Uriah Boyden quickly established himself as the most influential builder of turbines in America. Because he used a payment-by-improvement method for charging his many customers among New England's manufacturers, Boyden acquired as a result of exacting tests on his turbines a tremendous quantity of performance data, not to mention a huge personal fortune. This data was privately communicated to James B. Francis. At the time Francis was chief engineer to the Lowell Manufacturing Companies, who by 1849 had acquired a number of Boyden's patents and 'rights to use all his improvements relating to turbines and other hydraulic motors'. Moreover, they also held the patent rights in the Lowell area for Howd's turbines. So Francis, fully familiar with the work of Howd and Boyden (and Fourneyron for that matter), with the backing of Lowell's power hungry industries and with the facilities of the Lowell Machine Shop at his disposal, was well placed to make his own contribution to the development of water turbines.

His appraisal of the work of Howd and Boyden and the results of many tests conducted by himself suggested that the inward-flow system was the better arrangement. Just as Boyden before him had systematically and painstakingly worked out ways of designing the guide vanes and runner for a Fourneyron turbine, so Francis achieved the same for Howd's turbine. The prototype Francis turbines were unquestionably the most scientifically designed, the most thoroughly tested and the best constructed up to that time. Like Boyden, Francis was prepared to settle for nothing less than mechanical and hydraulic perfection. In a way it is a rather surprising development. The United States was at this time a new and expanding country, raw and undeveloped in many things, short of skilled craftsmen and with precious little tradition in engineering science. And yet it was here, with unprecedented scientific research and technical precision, that a key class of turbine was evolved.

In its original form the Francis turbine was hardly used; its importance lies in its triggering a process of development from inward-flow principles to mixed-flow, the evolution of the Francis turbine into its modern form. Safford and Hamilton have referred to the years after 1860 as the 'Cut and Try Period', when 'If a wheel did not come up to expectations, its buckets were chipped back, up or down, or its blades pounded, until it gave something better.' The United States went through the same – and it would appear inevitable – period as Fourneyron had touched off in France, namely a craze to try everything. American engineers excelled in this approach and, during the second half of the nineteenth century, an amazing variety of turbines and turbine accessories were built and tested in every imaginable combination. Even though the 'Cut and Try Period' presents quite a contrast to the strictly scientific approach of Boyden and Francis, a number of interesting developments in the end established themselves.

One of these innovations, elegant and important, was the diffuser. The use of a draft tube merely as a device to locate a turbine runner anything up to about 25 feet above tail-water level was an idea that Zebulon and Amasa Parker had patented in 1840 and Jonval and Koechlin had used in 1843. Uriah Boyden used it as well (Figure 44) and discovered that, suitably shaped and not necessarily vertically placed, a draft tube would function as a diffuser able to recover enough energy from the discharging flow to improve a turbine's overall performance by up to 3%. And then for the best part of twenty years, although wooden or sheet metal draft tubes were common, diffusers were scarcely used. In 1873, at the Holyoke test flume, the Risdon Company achieved good results with a draft tube diffuser. Subsequently developments in hydro-electricity, where every fraction of a percent of efficiency is critical, made the draft tube diffuser indispensable. Risdon's turbines also achieved a very high efficiency at part load. This latter consideration was an important factor in subsequent improvements. Once full gate efficiencies were regularly higher than 90%, it was part gate performance which claimed much of the attention.

The 'Hercules' turbine of 1878 and the 'Victor' of 1877 represent the first fully developed mixed-flow runners, for in these machines the flow was inward at inlet, axial through the body of the wheel and slightly outward at outlet. With greatly improved maximum efficiency and better all round performance at part load, the 'Hercules' and 'Victor' turbines and their derivatives were among the best hydraulic motors so far produced.

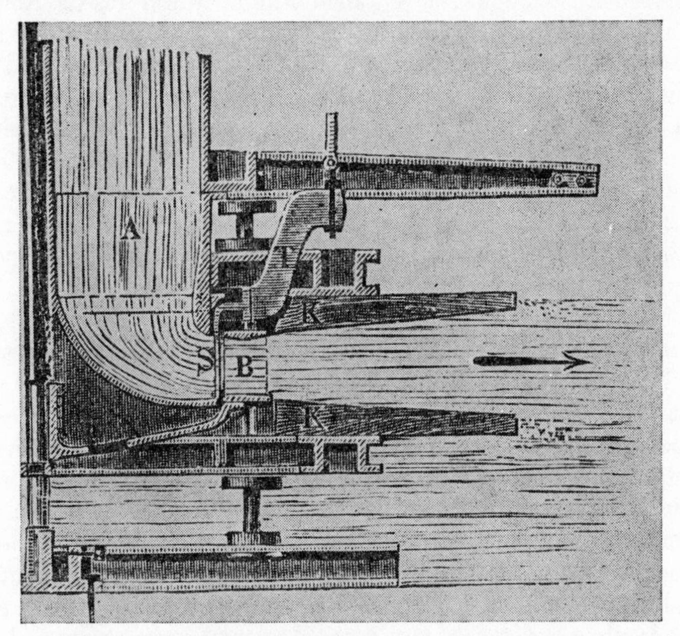

Figure 44 An early diffuser, set horizontally, to improve the efficiency of an outward-flow turbine.

Their most serious disadvantage, particularly in late nineteenth-century applications such as electrical power generation, was lack of shaft speed. Although the concept of specific speed was not used in the nineteenth century, it is interesting to note how in fact specific speeds slowly rose from a figure of 17 for the first Francis wheel, to the range 30–40 for the wheels of Swain, Leffel and Risdon and up to 48 for typical 'Hercules' and 'Victor' wheels. So for given heads and power outputs, higher and therefore more useful shaft speeds were becoming available.

Around 50, however, was by no means the highest specific speed obtained. Occasionally one finds such delightfully named machines as the 'Green Mountain' wheel, the 'Chase Special' wheel and the 'Austin' wheel achieving speeds of up to and occasionally over 100.[4] These specially developed low-head axial-flow wheels – in effect superior versions of the tub-wheel – were essentially low powered rural turbines and widely used in saw-mills and small workshops.

What has been said so far will have suggested that the United States might claim to have been the most influential contributor to turbine development in the nineteenth century. When the Pelton wheel is added to the list the claim is a substantial one.

Crude pressureless wheels certainly existed in North America in the first half of the nineteenth century but at what date their use became widespread is uncertain. The so-called 'Flutter' wheel was the American version of the Norse wheel and was as primitive a hydraulic motor as can be imagined. Much more interesting and rather better documented is the Rose wheel, whose origins can be dated at least as far back as 1840. This wheel represents a true ancestor of the pressureless turbine; twin jets were used to power a single row of curved metal buckets attached to one face of a circular disc mounted on a horizontal axle. The Rose wheel was apparently well known in New England and some northern States. Whether or not it penetrated as far as California where the most important developments were to occur, is uncertain.

Mining is always a key industry. Wherever the winning of metals, their ores or coal has been carried on in a big way, mining has always been central to extensive technological developments, at least in scale if not in degree of innovation. Coal-mining in Great Britain and metal-mining in Europe were pivotal industries making special demands on the engineer which were not infrequently met with radically new solutions.

Gold was discovered at Sutter's Mill in 1848. To win California's gold and its ores a full range of extraction techniques was applied: panning, fluming, digging shafts and drifts, and hydraulic methods. The extensive use of hydraulic mining meant the development of dams, pipelines, piping and hydraulic nozzles. The latter, called 'giants', were an especially important development because they constitute some of the earliest regular use of metal nozzles to form small diameter, high velocity jets supplied under very high heads. Other mining installations such as saw-mills, forges, grinding and crushing machines, the pumping requirements of shafts and fluming, all needed power. Everything was right, then, for the development of a hydraulic motor specially evolved to make use of the very high pressures which were 'on tap'.

The first water-wheel developed in the Californian gold-fields by the miners themselves was the hurdy-gurdy wheel. The view expressed by W. F. Durand, that it 'seems to have been an independent development in these mountains, a natural outgrowth of the environment and of the materials most conveniently at hand'[5] is eminently plausible. In a gold rush, or similar prospecting mania, many individuals with all manner of trades, skills and knowledge are thrown together against the same technological problems. The chances that their various talents will synthesize something new are always relatively high.

The hurdy-gurdy comprised triangular blocks of wood, about 4 inches thick, sandwiched between wooden shrouds which were attached to four or six spokes. The profile of the wheel's periphery has been likened to a circular rip-saw with the front face nearly radial (Figure 45). Clearly the hurdy-gurdy was cheap and easy to make while its performance and speed of rotation were adjustable, to a degree, by varying the jet size, the head and the wheel's diameter. In this form the hurdy-gurdy provided the bulk

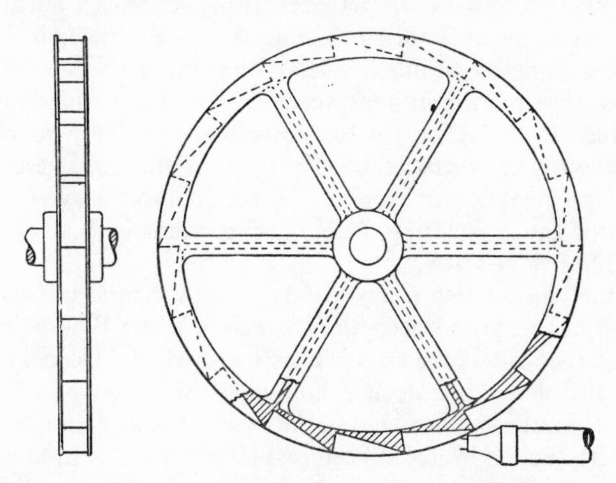

Figure 45 The simple construction of the hurdy-gurdy is evident in this illustration from the *Journal of the Franklin Institute* of 1895.

of power in the gold- and silver-fields of California and Nevada in the 1850s and early '60s.

It did not take long for the miners to see how to improve the hurdy-gurdy's efficiency, a particularly pressing need as more and more power was needed to drive stamp mills for ores which were becoming increasingly difficult to work. The hurdy-gurdy's buckets were in the form of a deep V. Without understanding the mechanics of the situation in any formal sense, hydro-power men in California saw that curved buckets would improve matters a good deal.

Around 1866 the Pacific Iron Works of San Francisco produced its first improved cast iron hurdy-gurdy for a stamp mill. The wheel's curved buckets were such a vast improvement that the old style hurdy-gurdy was a dead letter from that moment on. By the 1870s, the question of further improvement was very much in the air. Numerous people in the gold-fields, in engineering workshops, even at the University of California, were busy taking out patents, testing prototypes and formulating ideas relevant to 'California tangential water wheels with reaction buckets'.

Between them these early researches into bucket shapes suggested the three concepts which are basic to the modern pressureless turbine: tangential approach of jet to bucket surface; the double bucket with splitter which divides the jet into two halves; and the reversal of flow within the bucket so that the half-jets are deflected as nearly as possible through 180°. But lack of capital, lack of incentive and lack of initiative prevented anyone from making any impact commercially, until Lester Pelton came on the scene.

Pelton came to California in 1850, but failing to make his fortune in the gold-fields though he tried for fourteen years, he moved on to Nevada City and began manufacturing mining machinery including hurdy-gurdies. In common with other water-power improvers, Pelton in 1878 began his own search for an improved bucket shape. Various basic ideas were so much in the air that it has never been clear how much Pelton drew, consciously or otherwise, on the efforts of others. Nevertheless, he did carry out a great deal of experimental research for many years, and in 1880 patented the jet-splitting double bucket which is such a distinctive feature of the Pelton wheel.

The Pelton wheel began to be manufactured in San Francisco in 1880 and went from strength to strength. The Pelton Water Wheel Company'. catalogue of 1890 shows the astonishing rate at which business expanded in the first decade. Hundreds of Pelton wheels were sold all over North, Central and South America, and in Australia, South Africa, Japan and Europe. With 'Pelton Water Wheel' cast large on all his products (Plate 11), Pelton rapidly built up an international reputation, much to the chagrin of his competitors in California.

Pelton concentrated at first on rather small units of simple constructions His market, as he saw it initially, was the industries – mainly mining – of California, Nevada, Oregon and other western states. Their need was for something cheap, robust and transportable. Pelton wheels met all three requirements and yet provided high powers.[6] By 1900 the Pelton wheel was rapidly outgrowing its mining origins and elementary uses. An important element in the machine's reaching maturity was the influence of William A. Doble. Doble was introduced to Pelton wheels in the late 1880s when, as part of his general engineering work in California's mining districts, he noticed the one serious flaw in the design. The less than perfect shape of even the best buckets allowed turbulence and eddying which, in association with the sizeable concentrations of sand and silt present in most of the water supplies, was highly abrasive. Buckets were being cut to pieces in quite a short time. It took ten years for Doble to resolve the problem of designing a bucket around which the water-flow was uniformly smooth and turbulence free. Of his three patents taken out in 1899 the last shows the bucket shape which is 'modern'; subsequent developments have been minor. The 1899 buckets were so good that the Doble Company immediately went into the manufacture of Pelton wheels, much to the detriment of the Pelton Company.

Up to the early years of this century, the sole issue in the Pelton wheel's development was its buckets. The question of jet control had been very largely dodged by the simple expedient of deflecting the jet when it was not needed. In 1900 William Doble took out the first of many patents covering the control of jet nozzles by means of retractable needles. The upshot of Doble's excellent work on buckets and nozzles was the emergence of the Doble Co. as the makers of better Pelton wheels than the Pelton company. After more than a decade of commercial rivalry and numerous threatened if not actual patent litigations, the inevitable and sensible solution was reached four years after Lester Pelton's death. In 1912 the Pelton and Doble Companies amalgamated under the former's name.[7]

In 1900 two of the four basic modern turbine types had evolved: the Francis and Pelton turbines. The other types we have been discussing – Fourneyron, Jonval, Girard – were still far from redundant but their limitations were being increasingly highlighted by the superiority of the other two. Both were capable of high efficiencies – more than 90% – at full load, and generally greater than 80% right down to half load. Shaft speeds were commonly in the range 50–500 r.p.m. and hundreds of horse-power could be attained readily in a single unit. By 1903 Escher Wyss of Zurich had produced 8,200 h.p. from a single Pelton wheel and 10,000 h.p. from a single Francis turbine.[8] One other factor was emerging, namely the suitability of different turbines for different conditions and in particular different heads. It was generally accepted that for heads above 100 metres Pelton wheels were the best choice, while Francis turbines were well suited to the range 100–10 metres. Below 10 metres an inadequate machine was the double-discharge 'Samson' turbine but this was not much used outside the United States. The development of a high-power, high-speed turbine able to operate efficiently on very low heads, less than 10 metres, was to emerge as a pressing need in the twentieth century.

We must now look to see to what work the huge water horse-power being made available at the turn of the century was being applied. When one takes into account that by 1900 a single turbine maker, Escher Wyss, had built over 3,000 turbines capable between them of yielding over 500,000 h.p., it is perfectly obvious that the traditional areas of application are insufficient to account for these figures. It was in the field of electricity generation that the water turbine came into its own to the extent that today it is its almost exclusive function.

The development of electrical engineering was one of the dominant themes in nineteenth-century technology. Heavy electrical engineering was characterized by three critical issues: the problem of developing reliable and efficient generators and motors; working out ways to transmit electricity in large amounts over long distances, and in particular resolving that famous question of whether or not alternating or direct current was the better bet; and, lastly, there was the question of prime movers. Electricity is not a primary power source – it has to be generated, and in traditionally coal-burning areas, such as Great Britain, massive reciprocating steam engines were exploited to their limit before Sir Charles Parsons' elegant creation, the steam turbine, took over.

Hydraulic motors were coupled to electric generators at the very outset. Even in coal-rich Great Britain, a water turbine was used to light some very early Swan lamps at Cragside in Northumberland in 1880, and three years later hydro-electric power was used to drive two light railways in Northern Ireland. This region was particularly well endowed with hydro-electric potential and it was inevitable that some of the earliest British developments should take place there. It represents, too, the heyday of a local product and by far the most successful British turbine, the Thomson 'Vortex' turbine.[9] Patented in 1850 by James Thomson, the elder brother of Lord Kelvin, the 'Vortex' turbine was so named as a reference to its inward-flow lay-out. For its time the 'Vortex' turbine was an elegant design, two features in particular being notable: movable guide blades provided the best means of turbine flow control so far devised and the method became standard on inward-flow machines; and a spiral casing, soon to be dubbed 'scroll casing', was used to impart a proper direction and uniform velocity to the water as it approached the guide vanes. The first Thomson 'Vortex' turbines were in use in Ulster by 1852.

Initially it was inevitable that numerous hydro-electric installations should utilize natural falls of water. In various parts of France, northern Italy, Scotland, Austria, the Pyrenees and North America this approach marks the beginnings of the hydro-electric industry. Applications were often very limited especially in those numerous cases where, through shortage of money or lack of technical confidence or both the electricity was not transmitted. Metallurgical or chemical processes then had to be taken to the power. Aluminium extraction was a typical instance and this was the purpose of Scotland's first major hydro-electric scheme, the Foyers project on the east side of Loch Ness. These installations began work in 1896 and it is significant that both turbines and generators were imported; the Girard turbines came from France and the Oerlikon generators from Switzerland. Similar projects to Foyers were set up elsewhere in Scotland and in Wales early in the twentieth century.[10]

The full potential of hydro-electric sites, barring those of modest potential which were soon of only marginal interest anyway, depended on transmission of the power and distribution thereby over a wide area. Prior to 1895, for instance, the staggering quantities of energy available at Niagara had only been utilized to the smallest degree using hydro-mechanical methods. By the 1880s necessity and pride, combined with what can only be described as a self-conscious feeling that it was wasteful not to, stimulated American engineers and businessmen to devise a way of harnessing Niagara on a grand scale. From among several technical and commercial publications produced at the time, the following opening paragraph from a booklet published by the Business Men's Association of Niagara Falls, is interesting and typical:

> For many years it has been a matter of frequent comment, that at Niagara there existed an enormous water-power not utilized. Foreigners visiting the locality expressed their astonishment that a people so inventive and enterprising as the Americans should allow the unlimited power of Niagara to waste itself away without attempting to divert a fraction of the force flowing by their doors, to increase the material prosperity of their country.[11]

So began the epoch-making effort to harness Niagara, an enterprise which focused international attention on the role of hydro-electricity generally and the A/C versus D/C issue in particular. Actually the Niagara scheme was by no means the only major hydro-electric project afoot in the 1890s – there were several others in the United States, especially California, and quite a few in European countries such as Switzerland and Italy – but Niagara was by far the biggest in every sense. By 1896 there were ten 5,000 h.p. pressure turbines at work, all of the double-flow Fourneyron type. It is interesting that it was an early French type of turbine built in Switzerland which was chosen, rather than a more modern and more efficient American machine. The reason was essentially two-fold: in Europe there was greater experience of making big turbines, especially in the sphere of metal casting; and for Niagara's requirements of power, speed and head, the Fourneyron turbine, although relatively inefficient, was judged to be a more reliable and practical proposition. There can be no doubt that for turbine development itself Niagara was immediately important, for it stimulated tremendous efforts in the design and construction of bigger and better turbines. The Francis turbine in particular was greatly and rapidly improved in terms of speed, power and head; and this was to be of particular importance because Francis turbines performed so much better at part gate and were light on maintenance. Progress was made quickly. By 1901 Escher Wyss of Zurich were able to supply 5,500 h.p. Francis turbines running at 250 r.p.m. under 135 feet of water for Plant No. 2 of the Niagara Falls Power Company.

There is a sense, however, in which Niagara was, and still is, unusual; together with a few other sites, such as Lake Victoria and the Rhine Falls, nature provided both a fall of water and, very important, a steady flow all the year round. Such favoured sites apart, however, the hydro-electric engineer, in seeking to harness the world's hydro-power resources, had to contrive ways of creating high heads, storing large volumes of water and controlling flows seasonally and annually. In other words, dams were a crucial feature of hydro-electric work from the outset.

The earliest hydro-electric dams were part of run-of-river schemes, a logical starting point for many communities and industries because they were cheap and simple to set up, especially if a river dam suitable for conversion already existed. But in the final analysis, the river dam's low head and incapacity to regulate flow variations were serious limitations.

As Parts I and II of this book show, the nineteenth-century civil engineer had developed the art of dam-building for the formation of large reservoirs in the fields of irrigation and water-supply. Around 1900 one finds this accumulation of experience being increasingly applied to power production; the evolution of concrete dams in particular was a feature of the trend. The high dam in association with electric current transmission shaped a new concept of power production. The vast hydro-power reserves of remote and mountainous regions – the Alps, the Pyrenees, the Rocky Mountains – were now a realistic part of the world's energy resources. By 1900 the first steps to harness them were being taken. Since 1900 their development has been crucial and in the future may become more so.

15

Hydro-Electricity

IN THE SPECIFIC area of turbines the twentieth century has witnessed two basic developments: the introduction of the means to *design* turbines, and the emergence of new fundamental types; and the two are not unconnected. We have already seen that the nineteenth century slowly and with difficulty evolved the Francis turbine, a pressure machine, and the Pelton wheel, a pressureless turbine. The inherently complex hydro-dynamic characteristics of these motors took time to be appreciated. It is important to bear in mind that in the nineteenth century pressure and pressureless turbines were not designed on the basis of properly formulated theoretical principles. Indeed the highly confused discussion of turbine theory indicates that in many quarters there was the utmost misunderstanding of fundamentals. If anything, the evolution of the machines themselves was the most intrumental factor in elucidating and demonstrating the different natures of pressure and pressureless turbines.

By 1900 much had been resolved and experience, especially with large hydro-electric installations, indicated the pros and cons of various designs and the conditions for which each was most suitable. Still it is worth noting that the early turbine manufacturers were in business to sell, in the main, stock lines. Late nineteenth-century catalogues of manufacturers such as Peltons of San Francisco, Escher Wyss of Zurich and Gilkes of Kendal, advertised a range of turbines, one of which would be adequate for any given situation and application. This attitude prompted the remark that 'there has been too much effort directed towards adapting the plant to the wheels, rather than the wheels to the plant'.[1] It is a complaint which, in essence, was often repeated in the 1910s and 1920s. For modest jobs such an approach to 'turbine design' was not unreasonable, but when the first big hydro-electric developments came along so did the realization that for this type of work turbines would have to be 'one-off' creations closely designed for very specific conditions and requirements. The essence of the better informed and more thoughtful engineers' thinking on this question is well illustrated by the following extracts from a seminal paper by Chester W. Larner.[2]

> The writer is of the opinion that more trouble has resulted from the misapplication of turbines to conditions for which they are not suited, than from any other of the many mistakes which have been made in the equipment of water-power plants.
>
> . . . The cause of these mistakes is not far to seek. It lies chiefly in the failure to analyse the facts of these wheels and thus determine whether they are properly adapted to existing conditions.
>
> Comparatively few engineers realize to what an extent the performances of different types of turbines differ. The divergence is wide, and the speed selected determines largely the type of runner which must be used. In spite of this, the speed is usually determined from a consideration of generator economy alone, without regard to its effect on the economy of the turbine.
>
> The important point to be observed is that the wheel should be exactly suited to the conditions in order that the best results may be obtained.

So the twentieth century opened with the more perceptive engineers endeavouring to formulate a procedure whereby a turbine could be designed with characteristics as near perfect for the job as possible.

A crucial contribution to the field of turbine design was the introduction in 1903 by Professor R. Camerer of Munich of the concept of specific speed (N_s), a turbine parameter which is related to the three terms head (H feet), power (P h.p.) and shaft speed (N r.p.m.) by the equation $N_s = N.H^{-5/4}.P^{1/2}$. At first the significance and usefulness of the specific speed concept was misunderstood partly because of a confused terminology – it was variously referred to as specific speed, characteristic speed and unit speed – and also because its dimensions are not those of speed at all, either angular or linear.

However, experience in calculating and comparing the specific speeds of different machines quickly revealed the parameter's capacity to classify turbines of different types, pressureless Pelton wheels being distinguished by the lowest values of N_s, and Francis turbines by values in a higher range. It emerged also that between the two types there was a gap where a range of values of N_s between about 10 and 20 was not covered by any known turbine.

Specific speed concepts, in addition to serving as an index of turbine performance and a basis for comparison between different machines under similar conditions, also provided a means to relate different sized versions of the *same* machine: it was in short the means to reliable interpretation of model tests. Big hydro-electric turbines were (and are) an expensive proposition, so that an accurate prediction of performance was crucial. Thus it was that the model turbine began to assume an important role in design. Its use, however, was attended by a need to develop a very precise understanding of the hydrodynamic similarity between model and prototype if a valid design process was to be evolved. Model analysis has assumed great importance in the twentieth century. The acceptance of the turbine model was wholly symptomatic of the need to design accurately and confidently.

So far as the interpretation of model tests was concerned there were problems, notably with scale effect. The fact that the model is somewhat less efficient than the prototype does up to a point provide a useful safety margin. But as the scale ratio increases so does the scale effect, until a point is reached at which the discrepancy is so wide as to become a design hazard and correlation between model and prototype is no longer trustworthy. Ultimately, special correction formulae were evolved to deal with this problem.

Specific speed concepts not only stimulated a radically new approach to turbine design but also they awakened turbine designers to the need for new types of turbine and to the desirability of extending the specific speed range beyond the upper limit for Francis turbines which, at the time, was about 100.

High specific speeds, perhaps in excess of 100 on occasions, had unwittingly been obtained in the nineteenth century. But machines such as the Green Mountain and Chase Special, and for that matter the strange *turbine à hélice*[3] of the 1840s, were of very limited use and presumably went out of fashion without exerting any influence. All the same, though, they featured two fundamentally important and correct ideas – axial-flow and a very flat blade angle – both of which were crucial components in the evolution of the modern propeller turbine and its special variant, the Kaplan turbine.

So much effort and attention have been lavished on the development of propeller turbines that anything like a full history of the machine would require much space. We can merely observe here that by about 1910 research on both sides of the Atlantic was under way which was soon to bear fruit. Ideas for machines with fixed blades – in number anything between two and nine but usually four or six – were developed by various engineers and several designs went into production.[4] Notably there was the Lawaczeck turbine made under licence in a number of countries and, by 1920, a well-known machine.

These many and various fixed-bladed propeller turbines were not unsuccessful machines so long as they were not asked to accommodate varying conditions, especially part loads, which substantially reduced their efficiency. This restriction of the fixed-bladed propeller turbine, namely its high performance under very limited design conditions but poor performance elsewhere, was the fundamental issue quickly isolated and attacked by the most important contributor to the propeller turbine's evolution, Victor Kaplan.

At the beginning of the twentieth century, Germany was well to the fore in establishing experimental hydraulics laboratories. It was at the one founded at Brno in 1911 that Kaplan and his associates researched into the performance of propeller turbines and showed in 1913 how much the machine would be improved if it used variable pitch blades. Kaplan's innovations were so successful that efficiencies of greater than 80% were obtained throughout a percentage load range extending from $<50\%$ to $>100\%$.

Kaplan seems to have patented his ideas in 1913 but due, presumably, to the upheavals of the First World War no commercial installations were undertaken until 1918. The first Kaplan turbine was built by Ignez Storek for Hofbauer Witeve, Vlem, Austria and tested in 1919. Since it *was* the first of a new and very important breed of water turbines perhaps we should note its details:[5]

Diameter of runner	600 mm.
Power	35 h.p.
Head	3 metres
Discharge	1,100 lts./sec.
Speed	480 r.p.m.
Specific speed	700–800 (metric)
	157–179 (English)

Within a matter of three or four years these figures were improved out of all magnitude so that 1,000 h.p. wheels with metric specific speeds of 1,000 or more were readily available. A particularly instructive installation was at Lilla Edet in Sweden where a Kaplan turbine and two Lawaczeck propeller turbines of similar size and performance were operated under identical conditions. The diagram (Figure 46) shows very clearly the advantages of adjustable blades.[6]

All low head, high-discharge Kaplan and other propeller turbines had to be set with large draft tubes. High performance in Kaplan turbines depended critically on massive suction across the propeller and this highlighted further the problem of cavitation which had already proved itself a serious matter in pressure turbines using high efficiency diffusers. Typically, model analysis – from the outset an integral part of the Kaplan turbine's development – was used to investigate this troublesome and obscure issue, sometimes and graphically referred to as 'cold boiling'.

It is an inherent problem which continues to test the vigilance and judgement of turbine designers.

Some writers seem to see the Kaplan turbine (Plate 12) as a logical, indeed inevitable, final step in a chain of turbine developments stretching back to Fourneyron. Such phrases as 'subsequent developments have concentrated on refining and developing the basic Pelton, Francis and Kaplan turbines' are understandable but deceptive, and looking beyond hydraulic motors *per se* to the whole spectrum of hydro-power technology

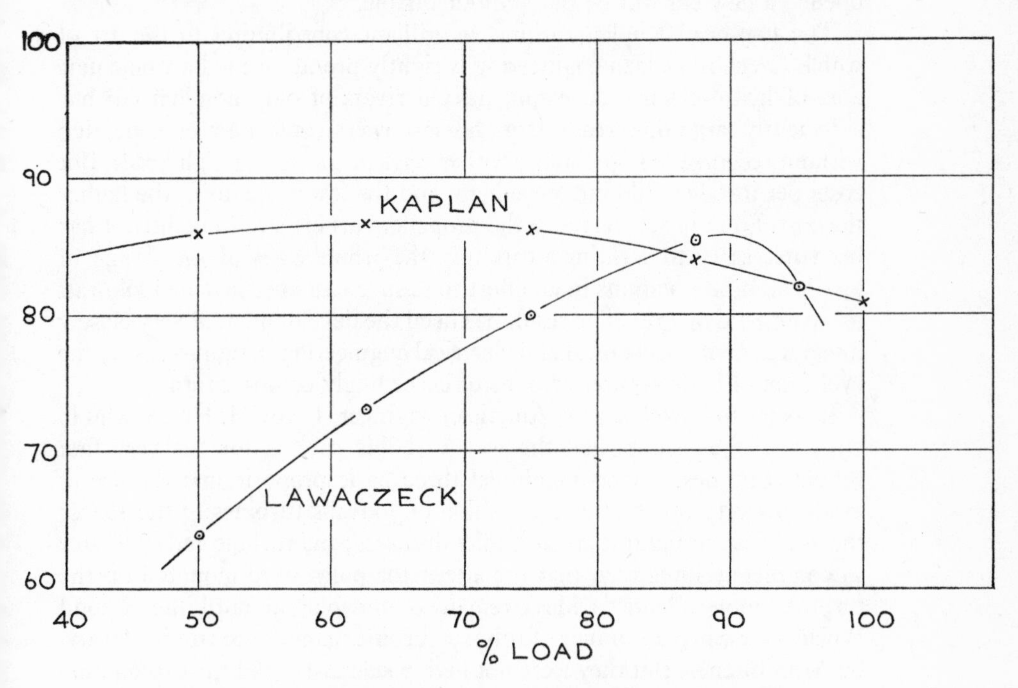

Figure 46 The comparative performances of the Kaplan and Lawaczeck turbines at Lilla Edet.

one is obliged to observe that the Kaplan turbine was in fact the threshold of a new era.

The Kaplan turbine itself has indeed been 'refined and developed', the principal effort having been directed towards the utilization of higher heads. Nowadays heads of 50 metres can be accommodated and for some small machines the figure is even higher. In an effort to realize Kaplan-like efficiencies at lower specific speeds, the principle of variable pitch blading has been extended to other machines. The very recent Dériaz turbine (Plate 13) covers a range of specific speeds similar to Francis turbines but with a significantly flatter efficiency curve. Unlike the Kaplan's, the Dériaz turbine's blades pivot on an axis inclined to the runner shaft rather than at right angles to it; the flow is inward and diagonal rather than axial.[7]

The Dériaz turbine is the most significant development in hydraulic prime movers of the last decade. Time will judge its ultimate importance but it is tempting to suggest that *the* key turbine development of this century is the use of variable pitch blades. Conceivably, as their performance range is extended, the Kaplan and Dériaz machines will comprise the whole repertoire of pressure turbines; and the notion that Kaplan's work opened a new era will be the more justifiable.

The low-head Kaplan turbine, a brilliant contribution to the art of which Czechoslovakian engineering is rightly proud, opened a whole new area of hydro-electric development. On rivers of only nominal fall but sufficiently large flow really large horse-powers could now be generated without recourse to high dams or other ways of contriving high heads. But costs per installed kilowatt were high, and the lower the head, the higher the cost. So in order to realize the propeller turbine's full potential it has been necessary to scrutinize carefully the whole basis of the design of low-head power stations in an effort to reduce costs per installed kilowatt to a competitive level. The result has been the development of very closely integrated civil, mechanical and electrical engineering design to ensure the evolution of hydro-power stations of exceedingly compact form.

It seems to have been the American engineer, Leroy H. Harza, who in 1919 and 1924 signposted the way with his patents for the very first tubular turbines.[8] Harza combined three basic propositions: the use of axial-flow very much after the fashion of Jonval's turbines of the 1840s; the use of the straightest possible inlet diffusers; and turbine and generator so completely integrated that the alternator poles were mounted on the turbine runner. Harza's ideas remained unexploited until the Second World War when a number of tubular turbine plants were run in Bavaria by Arno Fischer. But they were not highly successful and the exploitation of Harza's ideas to their fullest extent is a very recent achievement.

Nevertheless, the horizontal shaft axial-flow turbine, in forms other than those proposed by Harza, became a practical proposition somewhat earlier. The idea of transferring power from a horizontal turbine to a vertical generator through bevel gears was generally unsuccessful, and so the obvious was resorted to: the generator was placed close to the propeller, was direct axially driven and was submerged, a streamlined steel casing being used to enclose the electrical equipment and to provide access during operation. The bulb turbine, as it has come to be called, was first tried in Germany about 1940 but fell far short of expectations. After 1945 the focus shifted to France and the bulk of the bulb turbine's development has taken place there;[9] by the mid-1950s Electricité de France were testing bulb units of up to 320,000 h.p.

Probably the main interest of the bulb turbine centres on its role in modern efforts to realize the centuries-old dream of cheap, unlimited power from the tides. There is a great deal of energy in the tides but its exploitation is proving exceptionally difficult and promises to be highly expensive even at the most favourably endowed sites, which are not very many. Tidal power turbines have to cope with various problems: they must function efficiently at heads of only a few feet; they must be operable on both ebb and flow of the tide; and ideally it should be possible to run them as pumps as well as turbines. It was the bulb turbine which was chosen as the basis for French research in 1949 and after much development machines of this type were finally put to work in the Rance scheme in 1966. The Rance turbines must count among the most refined and flexible hydraulic machines so far constructed, for in addition to their turbining ability in the Kaplan manner, the generator can be motored so that the propeller then acts as a pump.

But even the evolution of this most elegant machine was by no means sufficient to stimulate wholesale exploitation of the tides. Schemes based on the River Severn, Morecambe Bay, the Bay of Fundy and the Gulf of San José, so often described, discussed and advocated, have equally often been shelved. And that is very largely where they lie at the moment. The French experience between Dinard and St Malo has proved that tidal electricity is not cheap and some critics believe that the world's first big tidal power station may ultimately be only one of a few.

Small-scale exploitation of tidal power has been carried on in the bays and inlets of northern Europe for hundreds of years (nor was the opportunity overlooked in North America). These small low-power tide-mills made little or no attempt to solve certain inherent weaknesses in all simple tidal installations;[10] the fact that power production is intermittent, that there is a variation in output between spring and neap tides and, most vexing of all, that whereas the availability of power follows a lunar cycle, demand follows a solar one. These are critical problems when one tries to develop tidal power on a big scale as part of a large electricity supply system. Somehow one must contrive the means to meet peak demand irrespective of the state of the tide. Over the past few decades techniques based on turbines doubling as pumps, single and double cycle systems and various twin-basin arrangements have been examined, but the best opinion seems convinced that the most satisfactory solution lies with pumped storage. This technique is one of the three most significant twentieth-century developments in hydro-power, the other two being the capacity to utilize very low heads and the use of turbines (i.e. the Kaplan and Dériaz types) and pumps with variable pitch blading.

While there can be no doubt that pumped storage has only become really important this century, its origins are older. Actually it has been suggested that pumped storage had its beginnings in the eighteenth century when water-wheel users, finding themselves short of water, installed pumps to recirculate rivers and streams. That procedure, however, represents a very debatable interpretation of the term 'pumped storage' as it is used today, and the true origins of the technique are really to be found in Europe at the end of the nineteenth century. The very first plant was built at Zurich in 1882 and the earliest installation to use centrifugal pumps was a textile mill at Creva-Luino on Lake Maggiore in 1894. Pumped storage was used for electricity generation at Ruppoldingen in Switzerland in 1904 and then at Schaffhausen. Soon after, in 1918, Britain's first venture into the pumped-storage field occurred on the River Tweed, an exercise which was not to be repeated for nearly forty years when the Sron Mor plant was put into operation by the North of Scotland Hydro-Electric Board in 1957.

By the 1920s pumped storage had proved itself in principle, a body of formal design theory and operating procedures was evolving, and the first installations in the United States were at work. In Europe pumped storage concepts had advanced to the point where two distinct types were acknowledged.[11] In mountainous areas such as the Alps, seasonal storage systems were introduced in which the pumping power was derived from spare hydro-electric output. At other places, notably in Germany, pumping on a daily cycle was required, and increasingly for this purpose off-peak steam generated electricity was used rather than the unrequired night-time hydro-electricity which had been favoured before.

After 1930 the desirability of extending the combination of base-load steam generation and peak-load hydro-generation was increasingly recognized. It has been as an element in such developments that pumped storage has been most influential to date. Early installations were dependent on massive and clumsy machinery comprising separate motors, generators, turbines and pumps, all mounted on a common shaft. For ease and economy of operation a pressing need was a hydraulic machine which would perform both as a pump or a turbine in conjunction with an electrical machine which could be run as a motor or generator. In fact the stimulus to the hydraulic engineer was considerable because pumped storage frequently requires very large flows to be pumped against considerable heads; for instance, as early as 1942 at Oberens in Switzerland the head was 1,000 metres.

That the principle of the pressure turbine is capable of reversal into a form of rotary pump, was appreciated in the nineteenth century. The inherent reversability of Thomson's vortex turbine was recognized in Osborne Reynolds' patent of 1875[12]; and centrifugal turbo-pumps were in use before 1900. Early in this century major manufacturers such as Escher-Wyss began to supply them. Mixed-flow pumps, i.e. the reverse of Francis' turbine, were developed in the 1920s along with the earliest axial-flow propeller pumps, originally proposed by P. K. Wood in 1896 and first used for irrigation and land drainage. In the 1930s there were two developments of some significance. In 1932 at the Baldeneysee power plant near Essen a reversible Kaplan pump-turbine was installed in a pumped storage installation. And then five years later the first of several reversible Francis machines were taken to Brazil for the Pedreira scheme. Subsequently in the 1950s the development of pump-turbines gathered considerable momentum as pumped-storage installations assumed greater economic importance.

The evolution of pumped-storage techniques in the last half century is symptomatic of a new level of hydro-power operations; it represents the third phase of water-power's development. In the first, the simple water-mill phase, the conversion of hydro-energy was immediate and local. In the second, the hydro-electric phase, the conversion was immediate but not restricted to local application. In the third phase, the energy is not only transmitted over a wide area but its production is also made to coincide with peak demand. Frequently, then, pumped storage plants represent something more than an altruistic quest for efficiency for its own sake; they reflect also that the hydro-power resources of a river, a region or a country have been so exploited and yet are so vital that the ultimate in operating efficiency from day-to-day, week-to-week and year-to-year must be achieved.

It is interesting to reflect on the varying forms of water-wheels and turbines, their changing patterns of use and the several factors which have shaped their evolution. In Figure 47 an attempt has been made to draw a genealogy from ancient times to the present day. One or two significant points stand out, notably how the evolution of hydraulic motors and their applications have ranged between the simple and the complex.

The horizontal wheel, initially a simple machine, achieved a complex form in the fifteenth and sixteenth centuries but is now ending its days in the same simple role in which it first appeared. The vertical wheel, simple in Vitruvius's time, was the centre of any number of new applications and some new ideas in the Middle Ages, then it settled into a routine role until the mid eighteenth century when there was a second burst of effort and initiative culminating in Poncelet's wheel. As such, the vertical wheel made a contribution to the evolution of Fourneyron's turbine. Otherwise the vertical water-wheel in its ultimate form – all-iron construction and pitch-back – was essentially a simple machine which Vitruvius would have understood.

The proto-turbine – the pre-Fourneyron phase – was immensely complex. Horizontal wheels, tub-wheels, reaction wheels of the Barker's mill and Euler types, and Poncelet's wheel, all influenced Fourneyron's work to a more or less degree, directly or indirectly. The turbines of Fourneyron and Jonval, a unifying and clarifying of many diverse and often muddled themes of the eighteenth and early nineteenth centuries, were themselves a prelude to a late nineteenth-century welter of activity and yet more muddled confusion and failure to get to grips with fundamentals. Niagara, and contemporary projects, were of great importance. Single turbines developing thousands of horse-power, with reliability and high efficiency at a premium, shifted the economic focus. The Francis turbine and Pelton wheel were quickly investigated, rationalized and standardized.

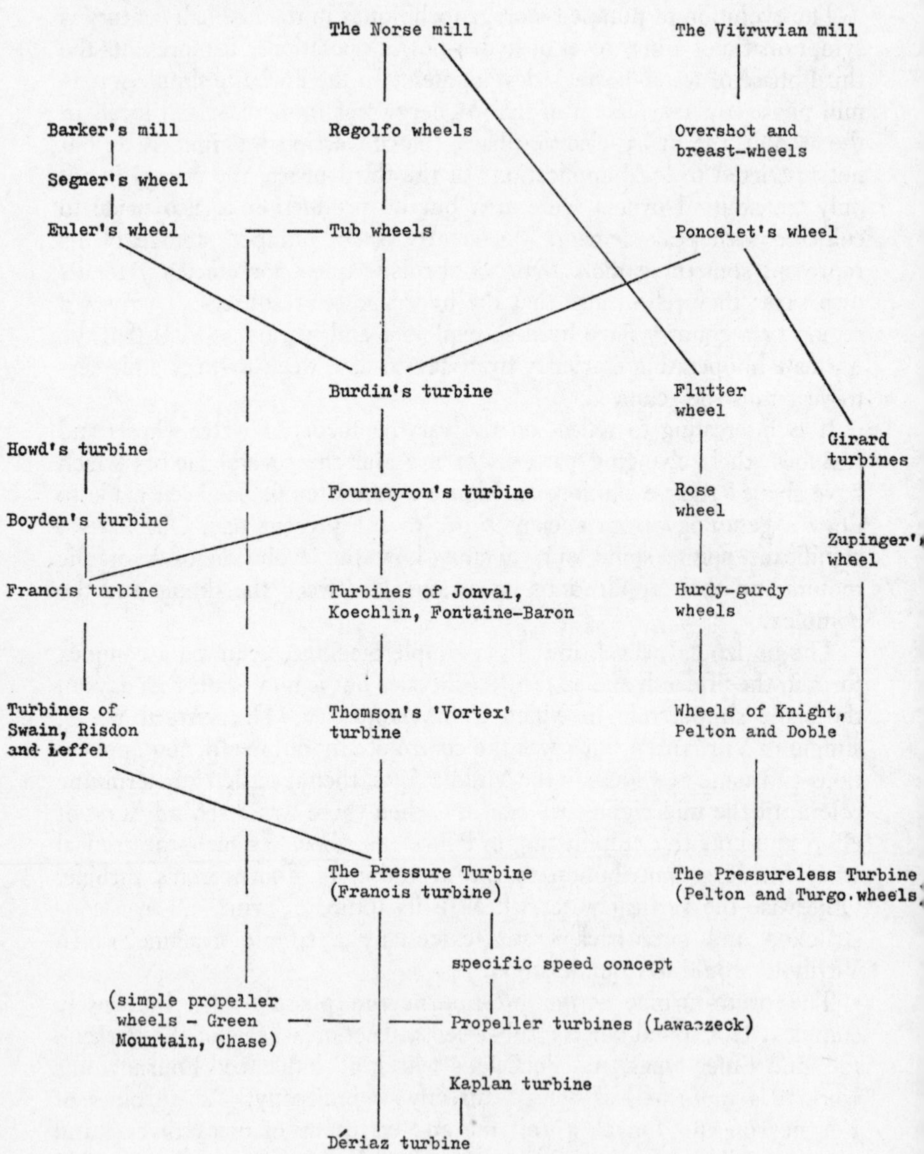

Figure 47 A genealogy of water-power.

The low-head turbine ran true to form. A simple proposition to begin with, the propeller turbine experienced a typical period of great variety of ideas and diverse experimentation, before it settled into the Kaplan form as the most suitable. Finally, the water turbine has merged with the water pump. The Kaplan and Dériaz machines in their pump-turbine form, while reflecting great refinement of technique and a vast accumulation of knowledge, have reduced the basic machinery of pumps and power to the barest minimum so far achieved.

It is notable too that water-power, while experiencing a period of bewildering ideas and many and various applications in the middle of its history (i.e. the Middle Ages to the nineteenth century), began as a very simple proposition – a means to grind corn – and the cycle has, in a sense, gone full-circle – now water-power generates electricity exclusively.

The history of water-power offers a unique opportunity to study the evolution of a power source over a very long period: Vitruvius's descriptions, remember, are about 2,000 years old. We have seen how the development of hydraulic motors responded to needs, how it shaped what men could do and, perhaps most interesting of all, how it resisted and adapted to its competitors. In the twentieth century, the exploitation of water-power has more and more been integrated with power generation from coal, gas and oil; and the latter have been by far the most prolific suppliers of the world's energy. At present about 2% of the world's power and about 25% of its electrical power is derived from water. So what of the future – can hydro-power maintain its 2,000-year-old record? In the next chapter we will consider this and other aspects of the present and future use of the world's water resources.

Part IV

Conclusion

16

Man and Water

THE STUDY OF the history of hydraulic engineering reveals two important characteristics which are prominent above any others: the fundamental importance of water throughout history and the continuous presence of certain basic technical, organizational, managerial and social themes.

A supply of water for drinking has been a primary need ever since urbanized communities were first formed. In many parts of the world the basis of successful agriculture continues to be irrigation, a technology whose ramifications extend well beyond the mere raising of crops. The application of water to power generation, once it had proved itself a viable proposition, was a powerful determinant in the evolution of several technologies in both east and west, and ultimately water-power exerted an important influence on the spread and development of industrialization.

All three aspects of hydraulic engineering are millennia old and in each case the fundamental concepts involved have not altered, as indeed they cannot. On the other hand the technology has been changed and improved radically especially in the last hundred years. The scale of modern operations is vast and complex by comparison with all previous works: the great irrigation schemes of antiquity, the water-supply systems of Roman and nineteenth-century cities, and the hydro-electric projects of the first half of the twentieth century.

However, hydro-technology *per se* is only one facet of the total relationship between man and water. In the past, as the body of this book shows, the development of methods to capture, control, distribute and use water has been the predominant issue. Building dams, manufacturing pumps and turbines, laying down pipelines and open channels, generating and transmitting electricity, and all the other work characteristic of hydraulic technology are now well within the compass of civil, mechanical and electrical engineering. However, notwithstanding its great scope and efficiency the content of the engineering repertoire is fast ceasing to be enough, and in many cases it is no longer the main consideration in hydro-developments at all.

Although the twin problems of providing adequate water quantities and ensuring their quality do pose significant challenges to engineering, they also raise even bigger political, economic, environmental and social issues. Already these are of considerable magnitude and they will undoubtedly become much more important. This shift away from specifically engineering problems towards those of water resources and their proper management represents the basic distinction between past and present hydro-technology. Formerly the basic issue was the provision of engineering works with which to exploit water-resources; today the problem is the provision of sufficient water to keep existing installations fully supplied and to make new projects feasible at all.

It is impossible to say what the *per capita* consumption of water was in ancient times, the Middle Ages or even the nineteenth century. Indeed, variations in modern statistics suggest that the calculation is not easy to make even now. Moreover, estimates become increasingly difficult when one attempts to calculate an overall *per capita* figure which includes agricultural, industrial and hydro-electric as well as domestic consumption. But without any doubt, the quantity is rising steadily the world over and sharply in the developed countries. Leading the field by a substantial margin is the United States whose each and every citizen is consuming and using something like 1,500 gallons of water daily.

This increasing *per capita* demand is placing an unprecedented strain on water resources, notwithstanding an increasing recourse to multiple usage and multi-purpose hydro-schemes whose potential is now being realized to a degree which was unknown and unnecessary in the past. The problem is being compounded by another and equally crucial factor: the rise in population. Ironically, the increasing demand for more water as a consequence of the growth in population is intimately linked with improved and enlarged supplies in the first place. One could expand at length on the role played by irrigation, water-supply and hydro-power in improving standards of living, increasing the global birth-rate and prolonging the expectancy of life. It is a story which is both heartening and depressing at the same time, a revealing view of man's capacity to improve and worsen his world simultaneously. However, two examples, the historical and technical context of which have already been introduced, must suffice; both represent a dilemma of classic proportions.

By 1900 the irrigation of India was a firmly established objective and for the most part its organization and administration was improving steadily. But at the same time a cycle of events, a chain of cause and effect, had been established which could never be reversed. Around 1850 India's population was about 175 million; by 1900 it was nearly 300 million; and now it is approaching 500 million. Irrigation has been a prime factor in this awesome increase which in turn demands more irrigation. Throughout the twentieth century India has built more and bigger dams, together with canal systems. The area now irrigated is about 80 million acres, say ten times what it was a century ago. And yet the food problem, the threat of starvation and the possibility of famine remain ever present. No one needs to be reminded that India has a population problem, but it does need to be said emphatically that the most elaborate and large-scale hydraulic engineering will not in the end be an answer to it.

Following its completion in 1902, the first Aswan dam brought such benefits to Egypt's agriculture and national economy that it was decided to heighten the structure by 35%.[1] Scruples over the preservation of the beautiful and ancient Temple of Philae, a factor in restricting the height of the first dam, were put aside. When the work was finished in 1912 the capacity of the Aswan reservoir had been more than doubled and a spiral began to form. In 1929 it was decided to raise the Aswan dam yet again, to a final height of 118 feet; and the reservoir's capacity doubled again.

In 1930, as a means to feed water into a more southerly canal system, the Nag Hammadi barrage was built between the Esna regulator (completed in 1909) and the dam at Assiut (see Figure 12, page 52). In the 1930s Nile utilization moved into a new and tricky phase and the problem of international co-operation loomed large. In order to develop its own irrigation, the Sudan had completed in 1926 the Gezira scheme, supplied with water from the Sennar dam on the Blue Nile. Once in operation the scheme was highly successful, and ways were evolved to ensure that the Sudan's use of Nile water in no way hampered Egyptian requirements.

But the spiral wound on. Believing that she had done her best to develop the Nile within her own frontiers, Egypt next examined the unorthodox concept of storing White Nile water in the Sudan to supplement storage at Aswan. Difficult negotiations were concluded in 1933 and the Gebel Awlia dam was commissioned in 1937. In 1940 the old Delta dams were finally pronounced unserviceable and replaced by the Mohammad Ali barrages cleverly designed and constructed to cope with the defect that had eventually undermined their predecessors, namely inadequate foundations laid on sand. Further Nile control and consequently more utilization was achieved when the Edfina barrage, basically a sea-sluice, was built in 1952.

The Nile was now controlled within the Sudan and Egypt by no less than nine dams and barrages of the storage and diversion type. But the Indian pattern had been repeated. A massive increase in the area irrigated, a wide range of crops under cultivation, and an increase in agricultural productivity had been matched by a quintupling of the population, from 5 million in 1870 to 25 million in 1970.[2] Once more hydraulic engineering was called upon to provide a solution, to yield not only more food but to give even greater insurance against all variations in the Nile's flow and to provide hydro-power as well. The Aswan dam concept was resurrected but the aim now was water storage on the most monumental scale behind a new dam.

The unhappy saga of the Saad el-Aali, the High Aswan dam, with its sinister political and ideological overtones, is still being unfolded and the final chapter is far from written. When the combined Russo-Egyptian engineering efforts of nine years were completed in 1969 a massive earth dam 365 feet high and 11,800 feet long was ready to impound the third largest reservoir in the world (130,000 acre-feet), capable in principle of irrigating 15% more land and providing 2,100 MW of electrical power. But the price was high and may go higher still.

An essential and cleverly contrived feature of the first Aswan dam was its multiple deep-level sluices designed to pass not just water but also the Nile's fertilizing silt. The High Aswan dam, partly for economic reasons and partly because of the technical difficulty of installing them, is not so equipped. This one omission has led to four important consequences each one as yet difficult to quantify with any precision, but threateningly present just the same. Lake Nasser is accumulating millions of tons of silt annually, deposits which will, in some finite time, choke the reservoir; fertilizer which was once available by a natural process now has to be either imported or manufactured in Egypt, using in effect the hydro-electric power developed at the dam itself; the Nile's failure to deliver silt to the Mediterranean has disturbed coastal equilibrium and initiated coastal retreat along the Delta; and by cutting off a natural food supply the sardine catch in the Eastern Mediterranean has declined from 18,000 tons to 500 tons per year.

These defects are bad enough, but others are significant as well. Evaporation is a problem. The hot, dry atmosphere of Nubia has a phenomenal capacity to evaporate water and estimates of around 10% have been put forward as the proportion of Nile water which is bound to be irretrievably drawn from Lake Nasser. Both bilharzia and malaria are water-borne diseases of long-standing concern in Egypt, in Africa and other parts of the world. Their incidence is being increased and not reduced by the 700-mile-long shore line of Lake Nasser and its miles of attendant irrigation canals where sluggish water and people innocent of the risks intermingle.[3]

Most mysterious of all Lake Nasser's defects is its apparent reluctance to fill. The reservoir was planned to reach capacity by 1970 but in June 1972 it was still less than half full. Evaporation and floods of less than average volume cannot account for this misbehaviour. A possible explanation of grave implications is the theory that the bulk of Lake Nasser's western bank is composed of porous Nubian sandstone, the edge of a 1 million square kilometre aquifer underlying the Libyan desert and capable of absorbing 3 billion cubic metres of water annually.[4]

The High Aswan dam has brought great benefits to Egypt, the many defects notwithstanding. The project is, nevertheless, a stage in the spiral, a phase in the inexorable process of supply and demand initiated a century ago with the first efforts at Nile control.

The demands of Egypt's increasing population and rising aspirations continue to grow and the High Aswan dam is not the final solution. But it might be the final effort of hydraulic engineers. Paul Ehrlich has claimed, probably pessimistically, that: 'The U.A.R. will have to complete the equivalent of four more Aswan dam projects in the next twenty-four years just to maintain its present inadequate level of nutrition.'[5] Hydraulic engineers for a hundred years have carried out some fine works of irrigation engineering on the Nile. But the impossible is impossible. When the Nile has been *totally* dammed and *fully* harnessed the contribution of hydraulic engineering will be at an end. The history of the Nile in Egypt is a classic case of one country's complete dependence on a large river and a total commitment to exploit that river mercilessly. The Colorado is being treated similarly and one is bound to note that the Tigris, Euphrates, Indus and others will not be spared either. The performance of Lake Nasser highlights a long list of defects from which no large reservoir can be entirely free.[6] Lakes Volta and Kariba in Africa, the Mangla reservoir in India, and Lakes Mead and Powell in the United States are just five of the world's largest reservoirs which exemplify some or all of the attendant problems of evaporation, salination, siltation, loss of land and health hazards. Other issues, more or less important according to location and climate, are the possibility that the massive superimposed weight of a large reservoir will set off seismic movements, the likelihood that traditional up-river fishing and other industries will be ruined and the almost inevitable need to evacuate communities of people, sometimes in large numbers. Whether or not large reservoirs will lead to serious microclimatic changes is obscure.

In view of these weighty objections and bearing in mind that suitable sites for big dams and large reservoirs are becoming increasingly hard to find, it is not altogether surprising that some experts advocate radically new approaches to irrigation. Georg Borgstrom has concluded that in the future reliance on traditional methods will become increasingly unproductive by comparison with past results and in relation to population increase.[7] Borgstrom also calls up the full range of objections to large-scale water storage which we have already noted and he proposes increased utilization of underground aquifers as a possible solution, but only in a fashion commensurate with natural rates of replenishment.

A particular concern of Borgstrom's is the question of irrigating arid lands at all. Notwithstanding past efforts in this direction, current pursuits of the same objective, and some notably successful outcomes, judged by prevailing criteria, Borgstrom argues that in the long run the concept is fundamentally unsound. He points out that only about one-fifth of the world's food is grown in irrigated regions and of these regions the bulk are located in humid and semi-arid areas, not the arid ones. The notion that food production can be significantly increased by 'making the deserts bloom is dubious from the basic point of view of water economy alone'. Evaporation losses, the basic cause of this unsound economy, mean that desert irrigation can require anything between ten and fifty thousand times the water needed in those humid regions where irrigation is at its most efficient. Borgstrom's final conclusion is that we would be wise to replace water storage and transport in regions unsuited to traditional irrigation in favour of food production and storage in more appropriate climes followed by food transport to populations in semi-arid and arid regions.

Something of the same propositions have been put forward by Professor L. Dudley Stamp in 'Dams and Deserts – are our concepts wrong?'[8] Although lack of water is the fundamental characteristic of arid lands, Professor Stamp does not accept that it is necessarily the fundamental problem. And in any case, even if it is, the construction of dams and canals is not the answer; he assembles the usual set of objections. If arid lands must be irrigated, he cites the example of Israel as indicating the proper basis for future developments. From the rivers Jordan and Yarkon in the north, large quantities of water are pumped in pipes – free of evaporation and pollution and carefully controlled in amount – to the new desert city of Beersheba and its environs. Professor Stamp extends this concept by suggesting the desirability of solving the problem of arid lands with people rather than irrigation. Indeed, he notices that in the past this was in fact the situation: in the Indus valley, in Sumer, Akkad, Babylon and Nineveh, in the Negev and in the Americas. He observes that in modern times the combination of all-the-year sunshine, no hard winters and air conditioning has prompted a growth rate of cities in Arizona, New Mexico and Nevada which is three times the national average.

To what extent the Borgstrom and Stamp propositions for arid lands will come about in the near or even distant future, is most difficult to analyse. Quite apart from the various technical problems posed, the ultimate exploitation of their ideas suggest economic and political difficulties of the sort which international co-operation has so far failed miserably to solve. Within territorial frontiers perhaps the opportunities are better.

Problems of international co-operation are equally central to the present and future development of hydro-electric power. The world's hydro-power resources are neither evenly nor appropriately distributed. The considerable potential of the River Congo, for instance, could not at present be anything like fully utilized in Central Africa. Whether or not the means can be worked out to develop the river on a large scale and 'export' the power over a wide area remains to be seen. On the other hand, the concept of selling power to South Africa is, or at least was, a feature of the Caborra-Bassa project at present under construction in Mozambique.

In other parts of the world international agreements on hydro-electric development assume a different form. Important power-producing rivers frequently cross one or more international frontiers: for instance, the Danube from Germany through Austria, Czechoslovakia, Hungary and Yugoslavia to Rumania and Bulgaria; the Columbia river from Canada into the United States; and the Duero from Spain to Portugal. Satisfactory if not ideal arrangements safeguard each party's interests and overall the rivers are efficiently and equitably utilized.

It is an interesting exercise to look back to the beginning of this century to the sort of predictions which were then being made about future energy resources. As replacements for the fossil fuels, much was expected from the direct conversion of solar energy, from a re-birth of wind-power and from the tides.[9] In fact it does not now seem that as producers of large scale power any of these three will prove important, certainly not in the near future. Predictions of half a century ago have been totally upset by the one power source which was then wholly unsuspected, nuclear power. It is in combination with this that hydro-power's future lies. There is much water-power on the planet yet to be harnessed, especially in Africa and South America, although the global appetite for kilowatts being what it is, and fast increasing, even the most lavish and extensive exploitation of hydro-power could meet but a small portion of the total demand. In any case there is a natural rate of loss of hydro-power potential as many existing installations find their reservoirs choking with silt. Notwithstanding this and other difficulties, the future of water-power is almost certain to be as a supplier of peak-load power in a system depending on nuclear energy for its base load. Indeed, from most environmental and pollutional standpoints the faster this state is reached the better.

It was noted earlier that in modern times questions of water quantities and purity have assumed great prominence, and nowhere is this more evident than in the field of water-supply.

Whenever raw water contains substantial quantities of natural, human, agricultural and industrial contaminants, its reduction to a level of purity which makes it safe to drink is complex and expensive. Moreover many existing water-treatment plants are by no means well equipped to deal with a wide range of the more obnoxious and intractable pollutants which have appeared in quantity over the last few years. Among such substances are phenols, 'hard' detergents, radio-activity and a variety of pesticides, fungicides and insecticides, all of which can and do find their way into surface- and ground-water sources. Current concerns over water pollution are not just a matter of principle, they are also a matter of money and a great deal of money at that.

Beyond the problems of the economics of purifying polluted water, however, stands the even more paramount question of water quantities. The history of water-supply has been, to a large extent, the history of people's search for more and more water, especially in the last two centuries. In certain areas, and Great Britain is no exception, increasing population numbers and/or rising standards of living have already reduced the margin between supply and demand to an uncomfortably low level. Unlike mineral, land, and certain energy resources, all of which are referred to as 'capital' resources, water, because it continuously regenerates itself and will never be used up, is conventionally and all too conveniently regarded as an 'income' resource. This is a seriously misleading view. The quantity of water circulating in the hydrologic cycle is fixed and has been for millennia. Is there a danger that we shall harness all of it? And if so, what then?

The quantity of water in the biosphere can be classified into various forms according to the following percentages.[10]

Oceans	97·61
Polar ice, glaciers	2·08
Ground-water	0·29
Lakes	0·017
Soil and subsoil moisture	0·005
Rivers	0·00009
Atmospheric water vapour	0·0009

Evidently the water resources which mankind has been so ruthlessly appropriating for several thousand years are but a tiny fraction of the planet's total reserves. The interesting question for the future is whether or not methods can be worked out which will enable regions deficient in water resources to tap the fresh water reserves of glaciers and the polar ice-caps and to convert sea water. To some extent the latter is already being done. In Aruba and Curaçao in the Netherlands Antilles, in Kuwait and in Bermuda desalination plants have been at work for some time 'manufacturing' drinking water from the only available source. The cost per gallon, however, is high, even when as in Kuwait the required energy input is cheap. This is the fundamental problem with desalination plants of all types and in so far as one resource problem is being solved at the expense of another, the concept is intrinsically of questionable merit. Conceivably, a proper objective for the contemplated increase in the use of solar power is fresh-water production. At present levels of solar technology however, the capital cost would be high and the space requirements formidable, about 1 square metre to yield 5 litres per day.

The use of polar ice is beset by many difficulties, not least the question of transportation. Nevertheless, a number of sober assessments[11] suggest that the large icebergs of Antarctica could be towed at reasonable cost to Australia and the western coasts of Africa and America. No one is taking up the idea at present, although one should bear in mind how acute the water problem already is in California, for instance. Stupendous plans to pipe water from Alaska have been projected, and so have immensely expensive schemes to desalt more than half a million gallons of sea water per day using nuclear power. That being the current level of thinking on the West Coast, the concept of 'importing' icebergs is by no means far-fetched.

Desalted sea-water (and conceivably melted polar-ice) represents a viable but at present very expensive solution to dwindling water resources for drinking and irrigation. Another possibility is the wholesale diversion of rivers which at present run to waste into those which are being utilized.[12] Fundamentally, such a concept represents an effort to rearrange an otherwise unsatisfactory distribution of hydraulic resources.

The Snowy Mountains scheme in Australia has demonstrated the possibilities of increased power production and extended irrigation by directing some of the flow of adjacent river basins into the Murray. In northern California a degree of southward diversion of northern-flowing rivers has already been achieved. Other proposed river diversions are on a massive scale, although technologically they are perfectly feasible. The diversion of Columbia river water into the Colorado would unquestionably lead to protracted and heated controversy. At least in the uninhabitable wastes of northern Canada, arctic Russia and central Brazil plans to divert such rivers as the MacKenzie, the Ob, Yenesy and Lena, and the Amazon have many attractions so far as resource utilization is concerned. Whether or not they make ecological sense is another matter.

There is little doubt that the world is facing a water problem the gravity and nature of which varies from place to place. Sometimes the crucial questions are of a very localized nature, more usually they are national issues, occasionally they assume the proportions of continental problems. The technology with which to achieve solutions to a large extent already exists. Whether or not society and its political and economic leaders have the determination to tackle the problem – assuming they are fully conscious of it, which is by no means certain – is another question. Reading through the considerable body of recently published articles, reports and books on water resources, pollution and hydro-technology, it is significant how widespread is the agreement on these points.

Michael Overman has written[13]:

> Technologically, therefore, man is equipped to meet the water shortage, but what is not clear is whether world governments and

authorities are bold enough to grasp the fuller implications of the problem. If man does fail to take the steps today that will supply him with ample fresh water tomorrow, the failure, in the final analysis, will be a failure of management, and of management alone.

In essence, and with variations in emphasis, this view represents an expert consensus which has yet to make the desired and necessary impact.

Notes

(Some of the bibliographical references are complete within the following Notes. For the remainder, please consult the Bibliography.)

Chapter 1 Ancient Irrigation

1. The topic is examined in, for instance, FORBES, Vol. II, pp. 1–22, J. W. GRUBER: 'Irrigation and Land Use in Ancient Mesopotamia' in *Agricultural History*, Vol. 21/22, 1947/8, pp. 69–77 and R. BRITTAIN, *Rivers and Man*, London 1958

2. The concept is explored in the article by COULBORN

3. See SMITH (1), Chapter 1

4. The problem is examined in T. JACOBSEN and R. M. ADAMS: 'Salt and Silt in Ancient Mesopotamian Agriculture' in *Science*, Vol. 128, 21 November 1958, pp. 1251–8

5. The Marib dam is described by SMITH (1), pp. 15–20 and R. L. BOWEN and F. P. ALBRIGHT, *Archaeological Discoveries in South Arabia*, Baltimore 1958, pp. 70–80

6. See the article by KEDAR and also N. GLUECK, *Rivers in the Desert*, New York 1959

7. An interesting development in Egypt was the use of river gauges. So called 'Nilometers' were numerous and of varying design and degree of elaboration; see BISWAS, Chapter 1

8. See Bibliography

9. Examples are GLICK, pp. 172–4, J. NEEDHAM: 'Science and Society in East and West' in *Centaurus*, Vol. 10, 1964–5, pp. 181–4, and E. R. LEACH: 'Hydraulic Society in Ceylon' in *Past and Present*, No. 15, April 1959, pp. 2–26

10. See DRUCKER, pp. 144–6

11. They are described in SMITH (1), pp. 35–8, VITA-FINZI and VITA-FINZI and BROGAN

12. See also the excellent book K. D. WHITE, *Roman Farming*, London 1970

13. A. G. DRACHMANN: 'The Screw of Archimedes' in *Actes du VIII Congrès International d'Histoire des Sciences*, 1956, Vol. III, pp. 940–3

14. M. L. KAMBANIS: 'The Drainage of Lake Copias' in *Bulletin de Correspondance Hellénique*, Vols. 16 and 17, 1892, pp. 121 and 322

15. The topic is discussed in K. D. WHITE, *Roman Farming*, London 1970, Chapter VI, in FORBES, Vol. II, pp. 41–6, and in WAGRET, pp. 51–3

16. More information is in SMITH (1), pp. 77–81 and ADAMS, pp. 77–8

17. For details of these works consult H. GOBLOT: 'Sur quelques barrages anciens et la genèse des barrages-voûtes' in *Revue d'Histoire des Sciences*, Tome XX, No. 2, April–June 1967, pp. 109–40 and A. K. S. LAMBTON, *Landlord and Peasant in Persia*, Oxford 1953

Chapter 2 Spain

1. An interesting discussion is GLICK, Chapters VIII and IX

2. GLICK, pp. 177–84

3. The edict is set out in full in SMITH (2), pp. 13–14; see too the section on Valencia in the same work

4. GLICK, Chapter XIII

5. SMITH (2)

6. An extensive treatment of the subject is R. L. BROHIER, *Ancient Irrigation Works in Ceylon*, 2 vols., Colombo 1934/36

7. A good account of the technique is M. D. COE: 'The Chinampas of Mexico' in *Scientific American*, Vol. 211, July 1964, pp. 90–8

Chapter 3 The Great Reclamations

1. For early Dutch and other northern European developments see WAGRET, pp. 54–74 and HARRIS (1), pp. 300–8

2. Discussed in F. STOKHUYZEN, *The Dutch Windmill*, London 1962

3. The performance of windmills is examined in R. L. HILLS, *Machines, Mills and Uncountable Costly Necessities*, Norwich 1967, pp. 32–3

4. A standard work on Fen drainage is H. C. DARBY, *The Draining of the Fens*, Cambridge 1956

5. For his life and work see HARRIS (2)

6. Vermuyden's life and work is studied in HARRIS (3)

7. See R. L. HILLS (1) for the history of wind- and steam-power in Fen drainage

8. A good account of the origins and early uses of centrifugal pumps is HARRIS (4)

9. For more details of the Italian story see HARRIS (1), pp. 308–15 and also the work by BAIRD-SMITH, Vol. 1

10. River engineering on the Arno is dealt with by PARSONS, Chapters XX and XXI

11. And apart from published works many reports were prepared by numerous engineers employed by different Italian cities and states, see PARSONS, Part V

12. GILLE (1), p. 181

13. On this point see the extremely interesting article by RETI (6)

14. Lupicini is discussed in PARSONS, pp. 347–57

15. The plates in question are Figures XCVII–CX in RAMELLI

16. Material on Castelli and others of the 'Italian school' can be found in HARRIS (1), pp. 314–15, ROUSE and INCE, pp. 59–72 and BISWAS, Chapters 8, 9 and 12

17. The issue is examined in S. LELIAVSKY: 'Historic Development of the Theory of the Flow of Water in Canals and Rivers' in *Engineer*, Vol. 191, 20 April 1951, Pt 1, pp. 466–8

Chapter 4 Colonial Irrigation

1. See for example the long list of Italian authors in BAIRD-SMITH, Vol. 1, pp. 301–4
2. J. de PASSA, *Voyage en Espagne dans les années 1816, 1817, 1818, 1819*, Paris 1823
3. See Bibliography for Capt. Baird-Smith
4. Sir C. R. MARKHAM, *Report on the Irrigation of Eastern Spain*, London 1867
5. C. C. SCOTT-MONCRIEFF, *Irrigation in Southern Europe*, London 1868
6. J. P. ROBERTS, *Irrigation in Spain*, London 1867
7. C. VINCENT, *Report of a Tour of Inspection of Irrigation Works in Southern France and Italy*, Madras 1882
8. W. H. HALL, *Irrigation Development: History, Customs, Laws and Administrative Systems Relating to Irrigation, Watercourses and Waters in France, Italy and Spain*, Sacramento 1886
9. In its published form the report is called 'Datos históricos acerca de todos los Pantanos construídos en España'. It appears in REVISTA DE OBRAS PUBLICAS, *Anales*, Tome I, 1896, pp. 5–18
10. W. J. M. RANKINE: 'Report on the Design and Construction of Masonry Dams' in *The Engineer*, Vol. 33, 1872, 5 January, pp. 1–2
11. For an excellent account of the history of dams on the Nile see ADDISON (2)
12. See WILLCOCKS (1)
13. They are described in WILLCOCKS (2); see also ADDISON (1)
14. On this issue and for further sources see SMITH (1), p. 238
15. FURON, p. 149

Chapter 5 The United States

1. For some discussion of the role of irrigation in the development of the West see the classic work by WEBB and also I. G. CLARK: 'Historical Framework' in *Aridity and Man*, edited by C. HODGE, Publication No. 74, American Association for the Advancement of Science, 1963, p. 87
2. A useful outline is O. W. ISRAELSON: 'The History of Irrigation in Utah' in *Civil Engineering*, Vol. 8, No. 10, October 1938, pp. 672–4
3. SCHUYLER, p. 164
4. These early arch dams are discussed in SMITH (1) and (2)
5. SCHUYLER, p. 164
6. A good account is L. A. B. WADE: 'Concrete and Masonry Dam-Construction in New South Wales' in *Min. Proc. Instn. Civ. Engrs.*, Vol. clxxviii, 1908–9, Pt iv, pp. 1–110
7. There were editions of WEGMANN in 1888, 1893, 1899, 1904, 1907, 1911, 1918 and 1927

8. CONDIT (2), p. 232
9. A useful outline is in ADDISON (1), Chapter 13
10. See Bibliography

Chapter 6 Wells and Springs, Qanats and Canals

1. For some discussion of this early phase try ROBINS and G. CLARK: 'Water in Antiquity' in *Antiquity*, Vol. XVIII, No. 69, March 1944, pp. 1–15
2. These topics are discussed in, for instance, FORBES, Vol. 1, pp. 145–89; C. E. N. BROMEHEAD: 'The Early History of Water-Supply' in *The Geographical Journal*, Vol. 99, 1942, pp. 142–51 and 183–96; and M. S. DROWER: 'Water-Supply, Irrigation, and Agriculture' in *A History of Technology*, ed. by C. SINGER et alia, Vol. I, Oxford 1954, pp. 520–57
3. An excellent modern account is H. E. WULFF: 'The Qanats of Iran' in *Scientific American*, Vol. 218, 1968, pp. 94–105
4. As for instance in the famous 'Joseph's Well' in Cairo. It is described in T. EWBANK, *A Descriptive and Historical Account of Hydraulic and Other Machines*, New York 1870, pp. 44–8. EWBANK's is a massive and frequently quoted work which, although a mine of information, is not without errors and needs to be used carefully. A recent and very valuable book is T. SCHIØLER, *Roman and Islamic Water-Lifting Wheels*, Odense 1973
5. The Samos tunnel has been studied by J. GOODFIELD and S. TOULMIN: 'How Was the Tunnel of Eupalinus Aligned?' in *Isis*, Vol. 56, No. 1, pp. 46–55
6. Inverted siphon is an unfortunate term for what is really a pipeline under pressure and what is *not* a true siphon in which the pressure is less than atmospheric. However we shall stick to conventional terminology
7. See FORBES, Vol. 1, pp. 151–2
8. The standard work is T. JACOBSEN and SETON LLOYD, *Sennacherib's Aqueduct at Jerwan*, Chicago 1935
9. See BUFFET and EVRARD, pp. 55–7
10. Later, however, we shall see that the Romans themselves were capable of massive investment in time, effort and lead to build some very large siphons
11. ROBINS, p. 54

Chapter 7 Roman Water-Supply

1. FRONTINUS, Book 1, Chapter 16
2. It should be emphasized that when the first serious archaeological work on Rome's aqueducts began about a century ago, there was much more to be seen than now – and access was much easier. Particularly since the Second World War what Bagnani called 'the sport of aqueduct trailing' has become much more difficult
3. Rome's aqueducts have a massive literature. Principal works in English are: ASHBY; E. van DEMAN, *The Building of the Roman Aqueducts*, Washington 1934; and E. M. WINSLOW, *A Libation to the Gods*, London 1963
4. FRONTINUS, Book II, Chapters 120 and 121
5. C. HERSCHEL, *Frontinus and the Water Supply of Ancient Rome*, Boston 1899, pp. 258–9

6. FRONTINUS, Book I, Chapters 7 and 9

7. The standard work on Lyon's aqueducts and siphons is MONTAUZAN. A more modern treatment is A. GRENIER

8. See the definitive work by CASADO

9. The matter is examined by J. P. MARTIN-CLETO, *El Abastecimiento Romano de Aguas a Toledo*, Toledo 1970 and briefly by J. A. GARCIA-DIEGO: 'Restoration of Technological Monuments in Spain' in *Technology and Culture*, Vol. 13, No. 3, pp. 426–9

10. Their details are in SMITH (2)

11. CASADO contains beautiful illustrations and drawings

12. Not even MONTAUZAN is free of defects, while ASHBY, pp. 34–7 is curious and confused and A. CHOISY, *Histoire de l'Architecture*, Vol. 1, Paris 1899, pp. 581–2 is idiotic

13. Air locking is not a problem in a pipeline under pressure, only when the flow rises above the hydraulic gradient and the pressure is reduced. The water pressure depends on the head operating; it cannot be reduced in any way or by any means

14. FRONTINUS, Book II, Chapter 65

Chapter 8 Fifteen Centuries of Neglect

1. A brief account is W. E. SMITH: 'Byzantine Aqueduct Still in Use' in *Civil Engineering*, Vol. 1, November 1931, pp. 1249–54. Also of interest is W. MATTHEWS, *Hydraulia*, London 1835, p. 233

2. This pollution question is mentioned by ROBINS, Chapters XII–XIV in three chapters which are useful on medieval water-supply generally

3. See for example J. A. DELMEGE, *Towards National Health*, London 1931, Chapter II, especially pp. 46–8

4. For this project see ROBINS, Chapter XVI

5. A standard and invaluable work on the history of London's water-supply is DICKINSON

6. The life and work of Myddleton is by GOUGH

7. A decided gap in the history of civil engineering is a study of practical surveying. We know something of the men and their instruments; we know little of how and how well they used the instruments

8. See D. R. HILL's beautiful and authoritative edition of al-Jazari's *The Book of Ingenious Mechanical Devices*, Dordrecht 1974, pp. 186–8 and 273–4

9. An attempt to set out the suction pump's history is S. SHAPIRO: 'The Origin of the Suction Pump' in *Technology and Culture*, Vol. V, No. 4, pp. 566–74; this is by no means the last word

10. According to J. BECKMANN, *A History of Inventions*, 2 Vols., London 1846, Vol. 2, p. 250, piston powered fire-fighting engines are recorded in Augsburg in 1518; the fact is consistent with pumped water-supply at the same time

11. For the history of the efforts of Juanelo and others at Toledo see L. RETI, *El Artificio de Juanelo en Toledo: Su Historia y su Tecnica*, Toledo 1967

12. BUFFET and EVRARD contains details and pictures

13. H. BEIGHTON: 'A Description of the Water Works at London Bridge' in *Phil. Trans. Roy. Soc.*, Vol. XXXVII, 1731, p. 5

14. Two valuable articles on Sorocold are: F. WILLIAMSON, 'George Sorocold of Derby' in *Journ. Derbyshire Arch. and Nat. Hist. Soc.*, Vol. 57, 1937, pp. 43–93 and R. JENKINS, *The Collected Papers of Rhys Jenkins*, Cambridge 1936, Chapter 20

15. J. FAREY, *A Treatise on the Steam Engine*, 2 Vols., London 1827 and Newton Abbot 1971. Steam-powered water pumping is described in Vol. 1, Chapters III and VIII; the use of steam engines in London is listed on p. 654

16. BAKER, Chapters I–IV

17. PARSONS, p. 241

18. Ramazzini on artesian flow is reproduced in F. D. ADAMS, *The Birth and Development of the Geological Sciences*, New York 1954, pp. 449–52

19. E. DARWIN: 'Of an Artificial Spring of Water' in *Phil. Trans. Roy. Soc.*, Vol. LXXV, 1785, pp. 1–7

Chapter 9 Nineteenth-Century Revolution

1. The Grenelle and other artesian wells are described by J. E. DUMBLETON, *Wells and Boreholes for Water Supply*, London 1953, Chapters I and VII and FURON, pp. 97–9

2. SMITH (1), p. 179. For an account of Thom's work see S. B. HAMILTON: 'Robert Thom and the Greenock Waterworks' in *Engineer*, Vol. CCXIX, 22 January 1965, pp. 168–71

3. Both the reports to which Telford contributed are given in full in T. TELFORD, *Life of Thomas Telford*, ed. by J. Rickman, London 1838

4. For Simpson's work and other early British filtration practice see the admirably detailed research in BAKER, Chapter V

5. This letter, well worth reading, is reproduced in L. PEARCE WILLIAMS, *The Selected Correspondence of Michael Faraday*, 2 Vols., Cambridge 1971, Vol. 2, p. 801

6. Useful details are in R. C. S. WALTERS, *The Nation's Water Supply*, London 1936, Parts IV and V

7. The Dale Dyke disaster is the subject of G. AMEY, *The Collapse of the Dale Dyke Dam 1864*, London 1974

8. An engineering account is G. H. HILL: 'The Thirlmere Works for the Water Supply of Manchester' in *Proc. Instn. Civ. Engrs.*, Vol. cxxvi, 1895–6, Part iv, pp. 1–25 and 70–127

9. The early history of dam design is in SMITH (1), pp. 192–207. See also Ref. 10, Chapter 4

10. See G. F. DEACON: 'The Vyrnwy Works for the Water-Supply of Liverpool' in *Proc. Instn. Civ. Engrs.*, Vol. cxxvi, 1895–6, Part iv, pp. 26–127

11. Ref. 10 above, pp. 29–49

Chapter 10 Other Countries, Another Century

1. Jervis' own account of the Croton scheme is in N. FITZSIMONS (ed.), *The Reminiscences of John B. Jervis*, Syracuse 1971, Chapters VII–X

2. See the excellent book by BLAKE

3. The classic account for New York is E. WEGMANN, *The Water-Supply of the City of New York, 1658–1895*, New York 1896

4. BLAKE, pp. 285–7

5. The history of coagulation techniques is in BAKER, Chapter XIII

6. There is nothing better than the detailed and definitive treatment in BAKER, Chapters XII–XXI

Chapter 11 The Water-Wheel

1. FORBES, Vol. II, pp. 78–9

2. The topic is analysed by E. C. CURWEN: 'The Problem of Early Water Mills' in *Antiquity*, No. 71, September 1944, pp. 130–46; by L. A. MORITZ, Chapter XVI; and in WHITE, Jnr, pp. 79–83

3. Both FORBES, Vol. II, pp. 95–102 and MORITZ, pp. 142–4 explore various factors which influenced the adoption of water-power. See also M. I. FINLEY: 'Technical Innovation and Economic Progress in the Ancient World' in *The Economic History Review*, Second Series, Vol. XVIII, No. 1, August 1965, pp. 35–7

4. PROCOPIUS, *De Bello Gothico*, Loeb Classical Library, London 1961, Book 1, Chapter 15

5. C. L. SAGUI: 'La Meunerie de Barbegal' in *Isis*, Vol. 38, Parts 3 and 4, February 1948, pp. 225–31

6. The origins of the horizontal water-wheel are assessed by: A. P. USHER, *A History of Mechanical Inventions*, Harvard 1962, Chapter VII; NEEDHAM, Vol. IV, Pt 2, pp. 366–9; WHITE, Jnr, p. 81; and CURWEN, as Ref. 2, pp. 134–43. See also WILSON (1)

7. It is interesting that the simple horizontal wheel with curved blades, known in France as *roue à cuillers* or *roue volant*, was sometimes called, especially in Languedoc, *rouet arabe*

8. Especially valuable for medieval water-power are GILLE (2) and FERRENDIER

9. GILLE (2), pp. 5–8

10. Such statistics are in FORBES, Vol. II, pp. 104–9 and GILLE (2), pp. 2–8

11. More details of the early applications of water-power are in GILLE (2), FERRENDIER and WHITE, Jnr

12. An excellent study is GILLE (1). The earliest known illustration of the horizontal water-wheel is in the manuscript of the anonymous German engineer of the Hussite Wars, *c.* 1430

13. See RETI (5)

14. This development is examined in SMITH (1), pp. 118–21 and p. 157

15. For details see DANILEVSKII

Chapter 12 The Industrial Revolution

1. D. S. L. CARDWELL: 'Power Technologies and the Advance of Science' in *Technology and Culture*, Vol. VI, No. 2, pp. 195–200

2. A. E. MUSSON and E. ROBINSON, *Science and Technology in the Industrial Revolution*, Manchester 1969, pp. 67–77

3. HILLS (2), Chapter 7

4. E. BAINES, *History of the Cotton Manufacture in Great Britain*, London 1835, p. 86

5. See WILSON (2) for details

6. The relevant part of Sir Anthony Fitzherbert's *Boke of Surveying* is quoted in R. BENNETT and J. ELTON, *Watermills and Windmills*, reprinted London 1973, pp. 190–1

7. SMEATON (1)

8. It is interesting that while Parent's work on water-wheels was less than adequate or accurate, it became well known and was often quoted, whereas his work on beam bending, containing substantially the best analysis of stress distributions up to that time, was hardly known at all. This reflects the relative degrees of prevailing interest in the two problems and does not support the notions that Parent's work was overlooked and ignored because his colleagues disliked him or because his books were printed in a minute format!

9. It is amazing how many writers persist in quoting 22% for Smeaton's results for undershot wheels. Smeaton's figures cover an efficiency range from 28–32%, average 30%. What one can do, and J. F. d'Aubuisson did it in his *Treatise on Hydraulics for the Use of Engineers* of 1858, is compile figures showing the *overall* efficiency of the *experiment*. These efficiencies will include energy losses in the water circulating system, especially the jet-forming sluice which on the scale of Smeaton's experiment was quite a small constriction. Smeaton, astutely and acutely aware of such scale effects, was at pains and some ingenuity to deal in *virtual* heads. In other words, he measured the velocity of the water at the wheel and then worked back to a virtual head. So he completely dodged the question of energy losses at the sluice and in the flume. Would that many later writers could see this and the need for it

10. Borda's work on water-wheels is reviewed in MASCART, pp. 98–117

11. SMEATON (2) p. 450

12. L. CARNOT, *Principes Fondamentaux de l'Équilibre et du Mouvement*, Paris 1803, p. 249

13. See Bibliography

Chapter 13 Early Turbines

1. See F. DI GIORGIO MARTINI, *Trattati di Architettura Ingegneria e Arte Militari*, 2 Vols., Milan 1967, Volume 1, pp. 141–59

2. Some are reproduced in RETI (2) and (4). For a general description of Turriano's manuscript see RETI (3)

3. See Bibliography

4. Consult NEEDHAM, Vol. 4, Pt 2, pp. 369–72 and 390–2

5. As shown for instance in RETI (1), Fig. 2

6. Illustrated in RETI (2), Figs. 1 and 2, and RETI (4)

7. The main contributions to the discussion are RETI (2) and (4) and WULFF

8. RETI (4), p. 390

9. D'AUBUISSON, p. 411

10. They are described in BELIDOR, Vol. I, Part 1, Book II, Chapter 1, pp. 302–4

11. D'AUBUISSON, pp. 409–10 and 416–17

12. L. EULER: 'Théorie plus complète des machines qui sont mises en mouvement par la réaction de l'eau' in *Mémoire de l'Académie de Berlin*, Tome X, 1754

13. A good discussion of Barker's mill and its nineteenth-century derivatives is in WILSON (3)

14. D. LANDES, *The Unbound Prometheus*, Cambridge 1970, pp. 181–2

15. Burdin's memoir was entitled *Des turbines hydrauliques ou machines rotatoires à grande vitesse*. Burdin's life and work has not been well studied by historians of engineering and by and large the same is true of the history of the water turbine generally. A frequently quoted but unreliable source is M. CROZET-FOURNEYRON, *Invention de la Turbine*, Paris 1924, written by a great-nephew of Benoît Fourneyron. Much more useful but disappointingly short is GILLE (3)

16. The most comprehensive and reliable account of Fourneyron's work is KEATOR

17. For more details see KEATOR, p. 300 and ARMENGAUD, pp. 276–301. Fourneyron's own works, listed by KEATOR, p. 301, are also very valuable but in fact he was by no means a prolific writer

18. The turbines of these designers are covered by ARMENGAUD, pp. 302–15 and 427–9. ARMENGAUD is generally a mine of information on a variety of turbines long since forgotten

19. Again ARMENGAUD is an essential source and also useful is J. WEISBACH, *A Manual of the Mechanics of Engineering*, 3 Vols., trans. A. J. du BOIS, New York 1877, Vol. 2, pp. 479–509

20. G. R. BODMER, *Hydraulic Motors*, London 1889, p. 36

21. Details of Girard's turbines with numerous illustrations can be found in G. RICHARD: 'Notes sur la construction et l'établissement des turbines' in *La Lumière Électrique*, 1883, Vol. VIII, pp. 25, 38, 74, 102, 139, 170, 204 and 232

Chapter 14 The Turbine in America

1. For this phase and later ones a prime source of information is SAFFORD and HAMILTON

2. FRANCIS, p. 2. FRANCIS is very important for the early history of the water turbine in America

3. In the discussion following SAFFORD and HAMILTON, Robert E. Horton argues Howd's case with vigour and at length

4. The specific speeds of these early high speed wheels has been the subject of some keen debate. See for instance F. NAGLER's contribution to SAFFORD and HAMILTON and also F. NAGLER: 'High-Speed Suction Turbines' in *Transactions*, ASCE, Vol. 89, 1926, pp. 648, 680 and 693

5. DURAND, p. 448

6. A portable Pelton wheel used in the Comstock Lode developed 125 h.p. but weighed a mere 220 lb, easily transportable by two men

7. Details of the manoeuvrings that went on are detailed by DURAND, pp.

450 and 514. Another dimension of the Pelton wheel's origins, namely whether or not it was a sufficiently novel invention to warrant the award of a medal by the Franklin Institute, is examined in *The Journal of the Franklin Institute*, Vol. CXL, September 1895, pp. 161–97

8. A very valuable source of information on nineteenth-century turbines is the early catalogues of manufacturers such as Escher Wyss. There is also *Escher Wyss, 1805–1955, 150 Years of Development*, 1955, an excellent company history

9. See WILSON (3), pp. 229–31

10. An account of early British hydro-electric developments is J. GUTHRIE BROWN: 'Sixty years of Hydro-electric Development in Great Britain' in *The Structural Engineer*, November 1956, pp. 373–403

11. *The Water-Power of the Falls of Niagara* published by the Business Men's Association of Niagara Falls, 1890, p. 9. There is of course a vast literature on the Niagara project, a standard work being E. D. ADAMS, *Niagara Power*, 2 Vols, Niagara Falls 1927

Chapter 15 Hydro-Electricity

1. LARNER, p. 314

2. LARNER, p. 313. Larner's paper and the considerable discussion which it provoked comprise something of a taking of stock, certainly so far as North America was concerned

3. See ARMENGAUD, p. 419

4. For a review of these machines see G. A. ORROK: 'High Specific Speed Turbines' in *Transactions*, ASCE, Vol. 89, 1926, pp. 616–24

5. From F. KNEIDL: 'Historical Note on the Kaplan Turbine' in *The Transactions of the First World Power Conference*, Vol. II, London 1924, pp. 503–6

6. Full details are in F. LAWACZECK: 'Large Low Head Water Power Developments' in *The Transactions of the First World Power Conference*, Vol. II, London 1924, p. 529

7. An account by the machine's creator is P. DERIAZ: 'The Mixed-Flow Variable-Pitch Pump-Turbine' in *Water Power*, February 1960, p. 49

8. M. GARIEL: 'Developments in Low-Head Hydro-Electric Stations' in *Engineering*, 4 November 1955, pp. 627–8. That Harza-like turbines may be important in tidal-power projects is discussed by E. M. WILSON: 'The Prospects for Tidal Power' in *Electrical Review*, 30 June 1967, p. 984

9. P. DANEL: 'The Hydraulic Turbine in Evolution' in *Proc. Instn. Mech. Engrs.*, Vol. 173, 1959, pp. 36–44

10. It is interesting that in his *Architecture Hydraulique*, Belidor gives some thought to the *theory* of tidal power systems. See BELIDOR, Vol. I, Part I, pp. 304–7

11. H. K. HAPPOLDT *et alia*: 'The Present State of Pumped Storage in Europe' in IEEE, *Transactions of Power Apparatus and Systems*, Vol. 82, 1963, p. 618

12. The essence of 'Specification of Patent No. 724' is given in O. REYNOLDS, *Papers on Mechanical and Physical Subjects*, 2 Vols, Cambridge 1900/1, Vol. 1, pp. 141–8

Chapter 16 Man and Water

1. For Nile dams see ADDISON (2)
2. ADDISON (1), p. 90
3. The issue is examined in R. H. LOWE-MACONNELL, *Man-Made Lakes*, London 1966, pp. 87–94
4. A. A. AHMED: 'An analytical study of the storage losses in the Nile Basin' in *Proc. Instn. Civ. Engrs.*, Vol. 17, October 1960, pp. 181–200
5. P. and A. EHRLICH, *Population, Resources, Environment*, San Francisco 1970, pp. 299–300
6. Aspects of these problems are examined in R. H. LOWE-MACONNELL, *Man-Made Lakes*, London 1966 and W. M. WARREN and N. RUBIN, *Dams in Africa*, London 1968
7. BORGSTROM, Chapter VIII
8. L. DUDLEY STAMP: 'Dams and Deserts – are our concepts wrong?' in *Jour. Instn. Elec. Engrs.*, April 1963, pp. 157–9
9. See E. W. GOLDING, *The Generation of Electricity by Wind Power*, London 1955, Chapter 2. Also relevant is E. AYRES and C. A. SCARLOTT, *Energy Sources*, New York 1952
10. From VALLENTYNE, p. 186
11. Such as HYLCKAMA, pp. 149–50 and OVERMAN, pp. 184–6
12. Briefly outlined in HYLCKAMA, p. 152 and OVERMAN, pp. 183–4
13. OVERMAN, p. 186

Bibliography

ADAMS, R. M. *Land Behind Baghdad*, Chicago 1965

ADDISON, H. (1) *Land, Water and Food*, London 1961
(2) *Sun and Shadow at Aswan*, London 1959

ARMENGAUD, Ainé *Traité Théorique et Practique des Moteurs Hydrauliques*, Paris 1858

ASHBY, Sir T. *The Aqueducts of Rome*, Oxford 1935

BAIRD-SMITH, R. *Italian Irrigation*, 2 vols, plus volume of maps and plans, London 1855

BAKER, M. N. *The Quest for Pure Water*, The American Water Works Association, Inc., 1948

BELIDOR, B. F. de *Architecture Hydraulique*, 4 vols, Paris 1737–53

BISWAS, A. K. *History of Hydrology*, Amsterdam 1970

BLAKE, N. M. *Water for the Cities*, New York 1956

BORGSTROM, G. *Too Many*, London 1969

BUFFET, B. and EVRARD, R. *L'Eau Potable à Travers les Ages*, Liège 1950

CASADO, C. F. *Acueductos Romanos en España*, Madrid 1972

CONDIT, C. (1) *American Building Art : the Nineteenth Century*, New York 1960
(2) *American Building Art : The Twentieth Century*, New York 1961

COULBORN, R. 'The Ancient River Valley Civilisations' in *New Perspectives in World History*, edited by S. H. Engle, Washington 1964

DANILEVSKII, V. V. *History of Hydro-engineering in Russia before the nineteenth century*, Israel Program for Scientific Translations, Jerusalem 1968

DARBY, H. C. *The Draining of the Fens*, Cambridge 1956

D'AUBUISSON, J. F. *A Treatise on Hydraulics*, translated by Joseph Bennett, New York 1858

DICKINSON, H. W. *Water Supply of Greater London*, Newcomen Society, London 1954

DRUCKER, P. F. 'The First Technological Revolution and Its Lessons' in *Technology and Culture*, Vol. VII, No. 2, pp. 143–51

DURAND, W. F. 'The Pelton Water Wheel' in *Mechanical Engineering*, Vol. 61, 1939, June, pp. 447–54; July, pp. 511–18.

FERRENDIER, M. 'Les Anciennes Utilisations de l'Eau' in *La Houille Blanche*, 1948 (Part 1, pp. 325–34; Part 2, pp. 497–508), 1949 (Part 3, pp. 121–33), 1950 (Part 4, pp. 769–87)

FORBES, R. J. *Studies in Ancient Technology*, 9 vols, Leiden 1955–64

FRANCIS, J. B. *Lowell Hydraulic Experiments*, New York 1909

FRONTINUS, *The Stratagems and Aqueducts of Ancient Rome*, Loeb Classical Library, London 1961

FURON, R. *The Problem of Water*, London 1967

GILLE, B. (1) *The Renaissance Engineers*, London 1966

(2) 'Le Moulin à eau' in *Techniques et Civilisations*, 1954, Vol. III, No. 1, pp. 1–15

(3) 'Les Sources Traditionelles d'Énergie' in *Histoire Générale des Techniques*, edited by M. Daumas, Vol. 3, Paris 1968

GLICK, T. *Irrigation Society in Medieval Valencia*, Harvard 1970

GOUGH, J. W. *Sir Hugh Myddleton*, Oxford 1964

GRENIER, A. *Manuel D'Archéologie Gallo-Romaine*, Vol. 4, Part 1, 'Aqueducs', Paris 1960

HARRIS, L. E. (1) 'Land Drainage and Reclamation' in *A History of Technology*, edited by C. Singer *et alia*, Vol. III, Oxford 1957, pp. 300–23

(2) *The Two Netherlanders*, Cambridge 1961

(3) *Vermuyden and the Fens*, London 1953

(4) 'Some Factors in the Early Development of the Centrifugal Pump' in *Transactions of the Newcomen Society*, Vol. XXVIII, pp. 187–202

HILLS, R. (1) *Machines, Mills and Uncountable Costly Necessities*, Norwich 1967

(2) *Power in the Industrial Revolution*, Manchester 1970

HERSCHEL, C. *Frontinus and the Water Supply of the City of Rome*, Boston 1899

HUNTER, L. C. 'The Living Past in the Appalachias of Europe: Water-Mills in Southern Europe' in *Technology and Culture*, Vol. 8, No. 4, pp. 446–66

HYLCKAMA, T. E. A. van 'Water Resources' in *Environment*, edited by William W. Murdoch, Stamford, Connecticut 1971

KEATOR, F. W. 'Benoît Fourneyron (1802–67)' in *Mechanical Engineering*, Vol. 61, 1939, April, pp. 295–301

KEDAR, Y. 'Water and Soil from the Desert: Some Ancient Agricultural Achievements in the Central Negev' in *The Geographical Journal*, Vol. 123, 1957, pp. 179–87

LARNER, C. 'Characteristics of Modern Hydraulic Turbines' in *Transactions*, ASCE, Vol. LXVI, 1910, pp. 306–86

MASCART, J. *La Vie et Les Travaux du Chevalier Jean-Charles de Borda*, Lyon 1919

MONTAUZAN, C. G. de *Les Aqueducs Antiques de Lyon*, Paris 1908

MORITZ, L. A. *Grain-mills and Flour in Classical Antiquity*, Oxford 1958

NEEDHAM, J. *Science and Civilisation in China*, 4 vols to date, Cambridge 1954–71

OVERMAN, M. *Water: Solutions to a Problem of Supply and Demand*, London 1968

PARSONS, W. B. *Engineers and Engineering in the Renaissance*, Baltimore 1939 (reprint 1967)

PONCELET, J. V. *Mémoire sur les Roues Hydrauliques à Aubes Courbes*, Metz 1827

RAMELLI, A. *Le Diverse et Artificiose Machine*, Paris 1588 (reprint 1970)

RETI, L. (1) 'Francesco di Giorgio Martini's Treatise on Engineering and Its Plagiarists' in *Technology and Culture*, Vol. IV, No. 3 pp. 287–98

(2) 'On Horizontal Waterwheels and Smelter Blowers in the Writings of

Leonardo da Vinci and Juanelo Turriano' in *Technology and Culture*, Vol. VI, No. 3 pp. 428–41
(3) 'The Codex of Juanelo Turriano (1500–1585)' in *Technology and Culture*, Vol. 8, No. 1 pp. 53–66
(4) 'On the Efficiency of Early Horizontal Waterwheels' in *Technology and Culture*, Vol. 8, No. 3 pp. 388–94
(5) 'The Problem of Prime Movers' in *Leonardo da Vinci – Technologist*, edited by L. Reti and B. Dibner, Burndy Library Publication No. 25, Norwalk 1969, pp. 63–96
(6) 'Leonardo and Ramelli' in *Technology and Culture*, Vol. 13, No. 4 pp. 577–605

ROBINS, F. W. *The Story of Water Supply*, Oxford 1946

ROUSE, H. and INCE, S. *History of Hydraulics*, New York 1957

SAFFORD, A. T. and HAMILTON, E. P. 'The American Mixed-Flow Turbine and Its Setting' in *Transactions*, ASCE, Vol. LXXXV, 1922, pp. 1237–56

SCHUYLER, J. D. *Reservoirs for Irrigation, Water-Power and Domestic Water-Supply*, New York 1901

SMEATON, J. (1) 'An Experimental Enquiry Concerning the Natural Powers of Water and Wind to turn Mills' in *Phil. Trans. Roy. Soc.*, Vol. LI, 1759, p. 100
(2) 'An Experimental Examination of the Quantity and Proportion of Mechanic Power necessary to be employed in giving Different Degrees of Velocity to Heavy Bodies' in *Phil. Trans. Roy. Soc.*, Vol. LXVI, 1776, p. 450

SMITH, N. A. F. (1) *A History of Dams*, Peter Davies, London 1971
(2) *The Heritage of Spanish Dams*, Madrid 1970

VALLENTYNE, J. R. 'Freshwater Supplies and Pollution: Effects of the Demophoric Explosion on Water and Man' in *The Environmental Future*, edited by N. Polunin, London 1972

VITA-FINZI, C. 'Roman Dams in Tripolitania' in *Antiquity*, Vol. XXXV, 1961, pp. 14–20 and plates II–IV

VITA-FINZI, C. and BROGAN, O. 'Roman dams on the Wadi Megenin' in *Libya Antiqua*, Vol. II, 1965, pp. 65–71 and plates XXI–XXVII

VITRUVIUS, *De Architectura*, Loeb Classical Library, 2 vols, London 1962

WAGRET, P. *Polderlands*, London 1968

WEBB, W. P. *The Great Plains*, New York 1931

WHITE, Jnr, L. *Medieval Technology and Social Change*, Oxford 1962

WILLCOCKS, Sir W. (1) *Egyptian Irrigation*, 2 vols, London 1913
(2) *The Irrigation of Mesopotamia*, London 1911

WILSON, P. N. (1) *Watermills with Horizontal Wheels*, published by The Society for the Protection of Ancient Buildings in 1960
(2) 'The Waterwheels of John Smeaton' in *Transactions of the Newcomen Society*, Vol. XXX, pp. 25–48
(3) 'Early Water Turbines in the United Kingdom' in *Transactions of the Newcomen Society*, Vol. XXXI, pp. 219–41

WITTFOGEL, K. *Oriental Despotism*, New Haven 1957

WULFF, H. E. 'A Postscript to Reti's Notes on Juanelo Turriano's Water Mills' in *Technology and Culture*, Vol. VII, No. 3 pp. 398–401

Glossary

Abbasid. A dynasty of 37 caliphs who were the titular rulers of the Islamic empire from A.D. 750 to 1258. The Abbasid line claimed descent from Muhammad's uncle, al-Abbas.

Acre-foot. The unit of reservoir volume commonly used in Great Britain and North America. It is a volume of water whose surface area is 1 acre and whose depth is 1 foot. 1 acre-foot is equivalent to 43,560 cubic feet, or 272,000 gallons, or 1,235 cubic metres.

Anicut. An Anglo-Indian word for a river-dam; especially characteristic of southern India.

Aqueduct. Strictly speaking, this is any artificial water channel. The term is often used, however, to describe what is really an aqueduct bridge.

Aquifer. An underground water-bearing stratum which can often be used as a source of water if tapped by a well or qanat.

Arch dam. A dam curved in plan and dependent on arch action for its strength; arch dams are thin structures and require less material than any other type.

Arched dam. Essentially a gravity dam which in plan view is curved. The name 'arch-gravity' dam is an alternative.

Atmosphere. A term widely used as a measure of pressure. One atmosphere is the pressure exerted by the atmosphere under normal conditions. It is equivalent to a column of water about 34 feet high and equals a pressure of 14·7 lbs/in^2.

Barrage. The French word for a dam or weir but commonly used in English for a large diversion dam, especially on rivers in Egypt and India.

Base-load. An electrical term referring to the steady round-the-clock output from a power station – usually steam generated.

Buttress dam. A special type in which a series of cantilevers, slabs, arches or domes form the water face of the dam and are supported on their air faces by a line of buttresses.

Cavitation. A hydraulic phenomenon characteristic of turbines, pumps and propellers. Air dissolved in the water comes out of solution at reduced pressure and sets up severe turbulence which can be very damaging.

Chinampas. The special type of cultivated plots built on the edge of the lakes around Mexico City. See Chapter 2, Note 7.

Cistern. An artificial reservoir used for drinking-water. Frequently they were built underground to keep the water clean and to prevent excessive evaporation.

Coagulation. The alteration of a solid that is dispersed or dissolved in a liquid into a coherent jelly-like or solid mass.

Corbelled arch. A form of arch made up from a series of horizontally placed stones which come closer together near the top of the arch and finally meet at the apex.

Core-wall. The central watertight wall of an earth or rock dam. Originally puddled clay was the material of construction; nowadays concrete is used.

Corvée. A period of unpaid labour exacted by the state from the peasant and lower classes. It was especially characteristic of French society before 1776.

Cusec. The standard abbreviation for cubic feet per second.

Cut-off trench. A deep excavation under the full length of a dam filled with an impervious material – puddled clay or concrete – designed to prevent or inhibit seepage under the structure.

Cylinder formula. The means to calculate stresses in a cylinder surrounded by, or full of, water. It has been used as an elementary technique for computing the stresses in arch dams.

Diffuser. The tapered discharge pipe below a water-turbine whose shape produces suction across the turbine wheel so increasing efficiency.

Diversion dam. A dam built across a river to divert water into a canal. It raises the level of a river but does not provide any storage volume.

Draft tube. A tube set below a water-turbine which allows the machine to be located up to 25 feet above tail water. Properly shaped a draft tube can also function as a diffuser.

Dyke. A word widely used in northern Europe for two different things. It can mean a ditch or channel used for drainage *or* a wall or bank used for flood protection.

Earth dam. A massive earthen bank with sloping faces and made watertight, or nearly so, with a core-wall and usually an impervious water face membrane.

Fascines. A long bundle of sticks, tightly bound, and used in flood-control works, river-bank reinforcement and temporary dams.

Fulling. A cloth-working process which cleanses and thickens the material.

Gin. Essentially a hoisting device driven by animals, usually horses, walking in a circle. It can be adapted to other purposes.

Gravity dam. A straight dam of masonry or concrete which resists the applied water load by means of its weight.

Grist-mill. A mill for grinding corn into flour.

Horizontal water-wheel. A water-wheel in which the wheel is mounted in a horizontal position on a vertical shaft; also known as a Greek or Norse mill.

Huerta. A Spanish word meaning a garden but often used to denote much larger irrigated areas along the banks of rivers.

Hydraulic gradient. A line joining the water-surfaces of the two tanks at each end of a siphon. Ordinates above the line measure hydraulic friction losses in the pipe; ordinates below the line are a measure of the pressure in the pipe. See Figure 17, page 76.

Hydraulic jump. The name of a phenomenon which converts fast, shallow flow to deep, slow flow in an open channel.

Hydraulic lime. A lime which will set hard in the absence of atmospheric carbon dioxide and is therefore valuable for submarine construction. Some hydraulic limes can be made from naturally occurring materials such as lias limestone and chalk marl. Otherwise non-hydraulic limes can be rendered hydraulic by the addition of natural materials such as pozzolana, Santorin earth or other volcanic materials; or by the addition of artificial ingredients such as burnt shale or clay, crushed tile or certain slags.

Inverted siphon. When a pipeline is used to carry water across a valley it is often called an inverted siphon. Whereas a true siphon operates below atmospheric pressure, an inverted siphon develops pressure higher than atmospheric.

Mere. A sheet of inland water or an area of water-logged or swampy ground.

Peak-load. The extra demand for electrical power which occurs at certain times over and above base-load. Peak demand is often conveniently met with water-power.

Penstock. A channel taking water from a reservoir to a water-wheel or, in modern times, a turbine.

Perennial irrigation. Irrigation which is practised all the year round. Different crops can be cultivated at various times in the calendar.

Pictograph. A pictorial sign or symbol; also writing made up of pictures.

Polder. A piece of land reclaimed from the sea or inland water. The term is used to describe a drained marsh, a reclaimed coastal zone or a lake dried out by pumping.

Prony brake. An early dynamometer – a machine to measure the performance of prime movers – developed in 1821 and frequently used thereafter to evaluate the output of water-turbines.

Putei. The vertical inspection and access shafts let into the underground sections of Roman aqueducts.

Qanat. A slightly sloping gallery driven into a hill so as to strike an aquifer; water will then flow from the aquifer down the qanat. Essentially it is a horizontal well.

Quintal. A French measure of weight equal to 100 kilograms.

Ringvaart. A circular canal around a reclaimed area of land into which drainage water is pumped.

Rock-fill dam. An embankment formed largely of dumped rock for stability and fitted with an impervious water-face membrane and core-wall.

Scouring gallery. A tunnel set low in a dam through which silt can be flushed by the pressure of a full reservoir.

Shaduf. A water-raising device consisting of a swinging beam on one end of which hangs a bucket and on the other end a counter-weight.

Specus. The Latin word for an aqueduct channel.

Transpiration. The process by which plants release water into the atmosphere from the surfaces of their leaves.

Umayyad. A dynasty of fourteen caliphs who ruled the Islamic world from Damascus between A.D. 661 and 750. The Umayyads were descended from the cousin of Muhammad's grandfather. When the Abbasids destroyed the Damascus caliphate, Abd-al-Rahman I escaped to Spain to re-establish the Umayyad line in Cordoba.

Uplift. Water percolating under a structure or into fissures can exert an upward force, potentially very dangerous, which is called uplift.

Vertical water-wheel. A water-wheel in which the wheel is mounted in a vertical position on a horizontal axle; also known as a Vitruvian mill.

Voussoir. One of the wedge-shaped stones which form a segmental, elliptical or parabolic arch.

Wadi. The Arabic word for a valley which becomes a watercourse in the rainy season. In Spanish the prefix 'guad' is derived from it.

Weir. A river dam used to raise the level so as to divert water into a canal or penstock.

Index